The Giant Canada Goose

The Giant Canada Goose

Harold C. Hanson

SOUTHERN ILLINOIS UNIVERSITY PRESS

CARBONDALE AND EDWARDSVILLE

Library of Congress Catalog Card Number 64–19798
Printed in the United States of America
Designed by Andor Braun

To Arlone, my wife,

who has helped me so much

PREFACE

AMERICA's rich heritage of birdlife has been marred by the extinction of some of its most unique and interesting forms. One of the subspecies believed to have been extinct for over three decades is the giant Canada goose, *Branta canadensis maxima*. The extinct status of this magnificent bird became more firmly established as various authorities successively reiterated this theme.

> It appears likely that a geographical race of large size once inhabited the northern prairies, for much heavier birds are recorded from this region. This breeding stock is now extirpated (Phillips and Lincoln, 1930:228).
>
> . . . also from Missouri (vicinity of Kansas City) in the same period [about 1920], was the report of the killing in older times of exceptionally large (i.e., 14–16 pound) geese of the Canada type, which were thought to represent a distinct, and possibly an extinct, race (McAtee, 1944:136).
>
> The giant Canada goose appears to be extinct (Delacour, 1954:163).
>
> The Giant Canada goose, which Delacour and Moffitt (Delacour, 1951) considered a distinct subspecies, is forever gone from the Mississippi Valley (Hochbaum, 1955:244).
>
> Now believed to be extinct (American Ornithologists' Checklist, 1957: 61).
>
> Occasionally smaller birds breed in this region still [the Great Plains], but it is probable that the original population [*B. c. maxima*] is extinct (Greenway, 1958:46).
>
> There can be no memorial to Missouri's goose [*B. c. maxima*], hypocritically dating the passing of the "last" of the race. We do not know when the last one fell (McKinley, 1961:7).
>
> The now extinct Giant Canada goose which sometimes weighed over 20 pounds was a prize even in those days (Boldt, 1961:13).

The original classic account of the giant Canada goose was written by William B. Mershon (1925), a copper mining executive of Saginaw, Michigan, based on his early acquaintance with the race in the 1880's in Kidder County, North Dakota.

> They were all big geese. We knew them as such. While having the general markings of the Canada goose, such as the black head and neck with white throat and cheeks, they were lighter colored, of a blue-ashy

general appearance, and the bodies were shaped differently — long ovals instead of round, chunky ovals [Fig. 1].

Even after the big flight came from the North and the white-fronted or laughing goose, the snow goose, the Canada and his smaller brother, the Hutchins, had come in thousands and thousands and were literally covering the fields at feeding time, these big geese held aloof and did not mix with the others. If they happened to alight on the same stubble field, it would be in a different part of the field from where the big lighting took place. They came into their feeding ground quietly, without any honking or noise; knew right where they wanted to go and did not circle about, but went right to their feeding ground.

They usually flew low, not so high in the air as the other geese. If we had a shot at a bunch of them, the geese we got were all big ones. We would never get out of the same flock one large one and one small one. By "small geese" I mean anything 12 pounds and under.

Happily, all reports of the demise of this magnificent goose, the largest in the world, have proved to be greatly exaggerated! The circumstances of its rediscovery, which have some amusing sidelights, should be briefly recorded.

For a number of years, on my return from fall fishing, I had observed the flock of Canada geese that winters in the city park of Rochester, Minnesota. I had always been perplexed by the size and coloration of these geese, but because nine months always had intervened since I had last seen the flock of Canada geese (*Branta canadensis interior*) at Horseshoe Lake in the southern tip of Illinois, I was never able to reach any firm conclusions about them. The history of the Rochester flock had indicated, however, that it was a unit independent of the main population of Canada geese in the Mississippi Valley Flyway. But the possibility that another large race might be wintering as a distinct entity in the heart of the migration route of the Mississippi Valley Flyway geese (*B. c. interior*) could scarcely be conceded.

The opportunity to solve this wildlife riddle was fortunately afforded me in mid-January, 1962, when I was invited by Forrest B. Lee of the Minnesota Department of Conservation to band, weigh, and measure a trapped sample of the Rochester flock. The initiation of these investigations was the result of joint planning by a group of Minnesota Department of Conservation and United States Fish and Wildlife Service personnel. On that memorable day, the temperature held around zero and a strong wind blew but this only added zest to the enterprise in which Forrest B. Lee, Robert Jessen, Thomas Hansen, and George Meyers of the Minnesota Department of Conservation, and Harvey K.

Nelson, Arthur S. Hawkins, and William J. Ellerbrock of the United States Bureau of Sport Fisheries and Wildlife participated. The work proceeded smoothly except for one hitch — we were obviously using faulty scales. The only question was "how faulty?" Therefore, we dispensed with further weighing until we could check the scales with some bulk food items of known weight. Arthur Hawkins responded to this fiscal challenge and purchased 5 pounds of sugar and 10 pounds of flour, but not before first weighing these items on the scale at the store. Upon our return to the banding site, a quick test of the scales revealed that the "impossible weights" we had been getting were correct. Now we knew beyond question that we were dealing with a very large race. But what race? The giant Canada goose had been repeatedly written into extinction and could not be a possibility. Only after returning home and checking Delacour's monograph, did I realize that the Rochester flock had to be *Branta canadensis maxima!*

This book, therefore, is not just a review of what is known of a supposedly extinct race of Canada geese, but is an attempt to summarize our knowledge of a race still moderately abundant in parts of the prairie provinces of Canada and in some areas of the United States. There are also many captive and wild flocks on state, federal, and private refuges in the northern plains and in the Mississippi Valley states. I have devoted considerable space to the taxonomy of the larger races of Canada geese because the burden of proof that the giant Canada goose still exists rests upon this study and because of the need for a definitive guide for identification of *Branta canadensis maxima* in the field. The origins of various refuge and private stocks of *B. c. maxima* are given in considerable detail as it is deemed important, insofar as it is possible, to establish the regional blood lines of these stocks.

During the preparation of this book, the question constantly arose as to how extensively I should summarize the biology of the giant Canada goose. I finally decided to include a fairly long chapter on behavior. My reason for doing this, and for amplifying other discussions, is that although the Canada goose is one of the most exciting birds for the general public, there has been no single reference work available that answers the questions that frequently arise about this species: At what age do they first nest? Do they mate for life? How long do they live? Which goose leads the flock in flight? Do they return to the same area each year to nest? These, and many kindred questions, can be answered only by a fairly thorough discussion of behavior. Fortunately, there was adequate literature on this subject to draw upon.

The physiology of wild geese is of unusual interest, particularly the physiological aspects of the molt. In this regard, my findings on the race of the Mississippi Valley Flyway, *Branta canadensis interior* (Hanson, 1962*a*), apply equally well to the other races of Canada geese. It therefore seemed appropriate to acquaint a wider audience with the high lights of some of the dramatic physiological changes that take place in Canada geese during the course of a year. In order that the interested nonprofessional may become familiar with methods used to determine the sex and age of Canada geese, I have taken the liberty of including in this book (Chapter 8) a recently published paper on characters of age and sex (Hanson, 1962*b*). It was deemed desirable to have this paper more widely available and in a more permanent form than originally published as the criteria of sex and age discussed have application to all species of geese.

After 20 years of research on wild geese, mostly on *Branta canadensis interior*, a definitive book on this race still remains one of my major objectives. However, before this can be accomplished, many studies in progress on the ecology, migration, morphology, endocrinology, physiology, biochemistry, and evolution must be completed. Until that time, I hope that this book will answer the general readers' questions about the larger races of the Canada goose, the "honkers," and particularly about this "new" addition to our avifauna.

Harold C. Hanson

Illinois Natural History Survey,
Urbana, *December 18, 1964*

ACKNOWLEDGMENTS

MY PRIMARY indebtedness is to Forrest B. Lee of the Minnesota Department of Conservation, who made it possible for me to study the Rochester flock in January, 1962, January, 1963, and in February, 1964. These investigations were further facilitated by the collaboration of biologists Robert L. Jessen, Thomas Hansen, Nicholas Gulden, and Leon Johnson of the Minnesota Department of Conservation, and Harvey W. Miller, Harvey K. Nelson, and William J. Ellerbrock of the United States Bureau of Sport Fisheries and Wildlife.

In April, 1962 and 1963, I visited various museums to study collections of skins and eggs of Canada geese. For courtesies extended to me, I am indebted to the following individuals and institutions: Earl W. Godfrey, National Museum of Canada (NMC), Ottawa; James L. Baillie, Jr., Royal Ontario Museum (ROM); Raymond A. Payntor, Museum of Comparative Zoology at Harvard (MCZ); S. Dillon Ripley, Peabody Museum of Natural History at Yale (PM); Dean Amadon, American Museum of Natural History (AMNH), New York; John W. Aldrich, National Museum of Natural History (NM), Washington, D.C.; James Bond, Philadelphia Academy of Natural Sciences (PANS); Kenneth C. Parkes, Carnegie Museum (CM); John L. Diedrich, Milwaukee Public Museum (MPM); Emmett R. Blake, Chicago Natural History Museum (CNHM); Paul W. Parmalee, Illinois State Museum (ISM); and Dwain W. Warner and Dennis Raveling, Minnesota Museum of Natural History (MMNH). I was also permitted to study two specimens of giant Canada geese collected by M. G. Vaiden at Rosedale, Mississippi.

In July, 1962, Harry G. Lumsden, of the Ontario Department of Lands and Forests, and I made an aerial survey of the breeding grounds of *B. c. maxima* in Manitoba. En route, landings were made on Chitek and Dog lakes and specimens were collected. The collaboration given by Mr. Lumsden on this flight and the use of an airplane piloted by David Croal and furnished by the Ontario Department of Lands and Forests are gratefully acknowledged. Additional reconnaissance flights over the breeding grounds of the giant Canada goose in Manitoba were made in July, 1963, and July, 1964, through the courtesy of the Game Branch of the Manitoba Department of Mines and Natural Resources. On these flights Eugene F. Bossenmaier was my informative guide and Frank Hanton was the pilot.

In September, 1962, with George C. Arthur, Illinois Department of Conservation, I trapped and studied flocks of giant Canada geese at three locations in Manitoba: the Delta Waterfowl Research Station, Delta; the East

Meadow Ranch, Oak Point; and at the Alf Hole Provincial Refuge, Rennie. For the support of this field program, I am indebted to Mr. Arthur, the Illinois Department of Conservation, and to Laurence R. Jahn and C. R. Guttermuth of the Wildlife Management Institute. Facilities, courtesies, and helpful advice were given in Manitoba by H. Albert Hochbaum, Director of the Delta Waterfowl Research Station; W. A. Murphy and George Kissack, owner and manager, respectively, of the East Meadow Ranch; and G. W. Malaher, Eugene F. Bossenmaier, Donald McEwen, Cal Ritchie, and Frederick W. Ans of the Manitoba Department of Mines and Natural Resources.

In October, 1962, Harvey K. Nelson and I visited refuges in Minnesota, North Dakota, and South Dakota to determine the subspecific identity of the flocks of Canada geese at the various refuges. For support of this field program, I am indebted to Mr. Nelson and the United States Bureau of Sport Fisheries and Wildlife. Information gained on this trip added immeasurably to my knowledge of the biology of the giant Canada goose. The courtesies and hospitality extended by Robley W. Hunt, Homer L. Bradley, John Dahl, Herbert H. Dill, Ned Peabody, Jerald Wilson, Merrill C. Hammond, James Salyer, John K. Bauman, Robert R. Johnson, David C. McGlauchlin, and Lyle J. Schoonover of the Bureau are deeply appreciated. At Jamestown, North Dakota, we were the guests of Carl E. Strutz who owns an outstanding flock of giant Canada geese. At the Lac Qui Parle State Game Refuge, Arlin C. Anderson showed us the flock of giant Canada geese which is being used by the Minnesota Department of Conservation in its propagation program to restore this race to west-central Minnesota.

A study of the flock of *B. c. maxima* at the Round Lake Waterfowl Station, at Round Lake in southwestern Minnesota, in December, 1962, was made possible through the courtesy of Kenneth Sather, Director, and the assistance of Glenn Smart, Biologist.

In February, 1963, the opportunity to visit the wintering grounds of the giant Canada in Rock and Washara counties, Wisconsin, was provided by Richard A. Hunt, James B. Hale, and Ralph Hopkins of the Wisconsin Conservation Department.

In early July, 1963, through the courtesies of Peter S. Suich and Glen A. Sherwood, I was able to examine a portion of the breeding flock of giant Canada geese at the Seney National Wildlife Refuge in the northern peninsula of Michigan. In late July, 1963, Donald W. Muir and William G. Leitch provided me with the opportunity to observe and photograph a portion of the flock of 22 giant Canadas on the Fort Whyte property of the Canada Cement Company.

In October, 1963, Dr. Alfred M. Bailey, Director of the Denver Museum of Natural History, made it possible for me to study the population of giant Canada geese in the Denver area.

In some respects this study had the aspect of assembling the widely scattered pieces of a jig-saw puzzle. That a reasonably coherent picture has been assembled is due very largely to the tremendous amount of information contributed by a host of waterfowl biologists. I have taken the liberty of quoting many of them at length.

For information on giant Canada geese in the prairie provinces of Canada, I am particularly indebted to Robert B. Klopman, Yale University; H. Albert Hochbaum, Delta Waterfowl Research Station; Alex Dzubin and E. Kuyt, Canadian Wildlife Service; Charles H. Lacy, Tom Sterling, Fred Sharp, Angus Shortt, and William G. Leitch, Ducks Unlimited (Canada); James L. Nelson, Saskatchewan Department of Natural Resources; Eugene F. Bossenmaier, Jack Howard, Donald McEwen, and Robert McWhorter (formerly), Manitoba Department of Mines and Natural Resources; R. Webb, Alberta Department of Lands and Forests; and Hamilton M. Laing, Comox, B.C.

Those who have contributed much helpful information on the status of the giant Canada goose in the United States are: Harvey K. Nelson, Arthur S. Hawkins, Harvey W. Miller, Robert R. Johnson, Herbert H. Dill, Leo M. Kirsch, David H. Shonk, and Cecil S. Williams, all of the United States Bureau of Sport Fisheries and Wildlife; Elwood G. Bizeau, Idaho Department of Fish and Game; Dale Witt, Montana Department of Fish and Game; Jack R. Grieb, Colorado Department of Game and Fish; Morris D. Johnson, North Dakota Game and Fish Department; the late Ray P. Hart, South Dakota Department of Game, Fish and Parks; George Schildman, Nebraska Game, Forestation and Parks Commission; Forrest B. Lee and Robert L. Jessen, Minnesota Department of Conservation; Richard A. Hunt, Wisconsin Conservation Department; George K. Brakhage, Missouri Conservation Commission; Miles D. Pirnie, Michigan State University; Karl E. Bednarik and Kenneth E. Allen, Ohio Department of Natural Resources; Carl E. Strutz, Jamestown, North Dakota; and M. G. Vaiden, Rosedale, Mississippi.

Specimens or heads of the large races of Canadas were collected for me for comparative study by Donald McEwen and Robert McWhorter (formerly) Manitoba Department of Mines and Resources; Harold H. Burgess, Allen E. Niemeyer, and Harvey W. Miller, United States Bureau of Sport Fisheries and Wildlife; Dale Horne and Clyde Scott, Idaho Department of Fish and Game; Vernon D. Stotts, Queenstown, Maryland; Richard W. Vaught, Missouri Conservation Commission; Glenn Smart, Round Lake Waterfowl Station; Paul W. Parmalee, Illinois State Museum; and Alfred M. Bailey and Jack Putnam of the Denver Natural History Museum.

The following have kindly contributed photographs: Roger W. Balham, Victoria University, Wellington, New Zealand; Harry G. Lumsden; Kenneth Brossmann, Oakes, North Dakota; William Brown, Cairo, Illinois; Morris D. Johnson; Ed Gross, Kenmare, North Dakota; Robert L. Jessen; David C. McGlauchlin, James Thompson, and C. J. Henry, United States Bureau of Sport Fisheries and Wildlife; R. A. Brown, St. Joseph, Missouri; Charles F. Yocum, Humboldt State College, California; Allan D. Cruickshank, National

Audubon Society; Robert Dodds, Ducks Unlimited, Winnipeg, Manitoba; Donald Wooldridge, Missouri Conservation Commission, and Gordon Yeager, *Rochester* (Minn.) *Post-Bulletin*. Photographs of the developmental stages of the giant Canada geese were taken by William E. Clark, formerly staff photographer of the Illinois Natural History Survey. I am also indebted to Wilmer Zehr and Marguerite Verley of the Illinois Natural History Survey, Section of Publications, for assistance in preparing several of the figures for publication.

A number of key references were supplied by Harry G. Lumsden, Forrest B. Lee, Richard A. Hunt; by Robert A. McCabe, and A. W. Schorger of the University of Wisconsin; and by Alex Dzubin.

Many invaluable comparisons of *B. c. maxima* with *B. c. interior* were possible as a result of support provided since 1946 by the Arctic Institute of North America and the Ontario Department of Lands and Forests for field work on Canada geese (*B. c. interior*) in the Hudson Bay Lowlands of northern Ontario.

A cornerstone of the information on which this book is based is the correspondence files of the late William B. Mershon and Ray P. Holland, thoughtfully made available to me by the latter. It is hoped that this book will also constitute a tribute to the dedicated efforts of Mr. Holland and the late William B. Mershon toward achieving the recognition and conservation of the giant Canada goose.

The writer is especially grateful to Dr. Thomas G. Scott, formerly Head of the Section of Wildlife Research, whose encouragement for so many years did so much to insure the continuity and success of my investigations of Canada geese, and to Dr. Harlow B. Mills, Chief, Illinois Natural History Survey, who gave me the freedom to pursue these studies on a flyway basis.

In the preparation of this book, the secretarial and editorial assistance of Arlone Kruegel Hanson is gratefully acknowledged.

I have benefited from reviews of portions of the manuscript by Harvey K. Nelson, Arthur S. Hawkins, Forrest B. Lee, Alex Dzubin, and H. Albert Hochbaum. My colleagues in the Section of Wildlife Research of the Illinois Natural History Survey made helpful suggestions regarding Chapter 13 and Dennis G. Raveling on Chapter 12.

Funds to aid in the publication of this report were received from the Wildlife Management Institute and the Rob and Bessie Welder Wildlife Foundation. For this support I am grateful, and wish to thank C. R. Gutter-muth and Clarence C. Cottam, Vice-President and Director, respectively.

To sum up, seldom have so many contributed so much to a single study. I am deeply appreciative.

CONTENTS

LIST OF FIGURES

LIST OF TABLES

LIST OF MAPS AND GRAPHS

MAPS

GRAPHS

The Giant Canada Goose

History of Discovery I

THE DISCOVERY and scientific recognition of most animals have proceeded in an orderly manner. With a few, however, the special circumstances surrounding their recognition have been associated with a degree of mystery, unreceptive officialdom, and general disbelief at the technical level. To some extent, the recognition and rediscovery of the giant Canada goose have encompassed all of these elements. It is a story of the Indian, and often the hunter and layman, having a more acute understanding of a bird than the professional. It is also a story of some poignancy and, as a tributary story of the "winning of the West" and the decimation of its flora and fauna, it is a bit of Americana. Because the problems of threatened species will always be with us as long as human populations continue to expand, the history of the discovery and near extirpation of the giant Canada goose is of interest and importance.

In a series of brief unsigned articles beginning in 1919, Ray P. Holland, then President of the American Game Protective and Propagation Association, raised the question, "How much does a goose weigh?" (*see* Holland, 1922). Although he reported that he had not weighed a Canada goose over 12 pounds, he sensed from the stories he had heard about big geese taken on the Great Plains (Fig. 2) that the ornithologists of the day had not by any means described all the races of large Canada geese that existed. Indeed, only one large race, *B. c. canadensis*, was recognized at that time.

One of the noted sportsmen of that day was William B. Mershon. His hunting experiences dated back to the 1880's, the high lights of which he summarized in a book in 1923. On February 24, 1922, Holland initiated correspondence with Mershon regarding the existence of a big goose and, in doing so, he began an exchange of letters with Mershon, other hunters, scientists, and administrators that lasted over a period of 17 years. It terminated at the death of Mershon in 1939 at the age of 84.

A sampling of a few of the letters received by Mershon and Holland from various individuals will chronicle the discovery and history of this race. In reading these extracts the reader must bear in mind the status of our knowledge of waterfowl in the 1920's and, particularly, that only in the past two years has our knowledge of the taxonomy and distribution of Canada geese begun to coincide with reality. The proneness of sportsmen to misjudge the size and weight of the game animals they bag has always been sufficient reason for scientists to view with scepticism reports of individual animals that departed widely from the *known* norm for a given species. Having failed repeatedly myself to recognize the authenticity and significance of reports of large Canada geese, it is surely not my intention in the following quotes to cast aspersions as to the scientific astuteness of the distinguished ornithologists who were consulted by Mershon and Holland in the course of their efforts to secure scientific recognition of the giant Canada goose. Rather, my only purpose here is to document the history of the efforts made to gain recognition of this race. Unfortunately, it is a story from which professional biologists and conservationists can take little satisfaction. But it is also not without its moments of humor.

R. P. Holland to W. B. Mershon, February 24, 1922:

I have been working on the theory that there is a distinct species of big goose for the last ten years, but have never gotten very far with it. I have a pile of data here in the office, but nothing conclusive enough to convince the ornithologists that such a bird exists.

Early in the inquiry, experts of the day became involved in the controversy.

George Bird Grinnell to W. B. Mershon on March 8, 1922:

If you get the big goose from Pettibone [a resident of Dawson, Kidder County, North Dakota], I think I would send it perhaps to [A. K.] Fisher at the Biological Survey. He thinks Dr. [Alexander] Wetmore could decide this question as well as anybody; and very likely he is as good a man as could be found.

R. P. Holland to W. B. Mershon on June 26, 1922:

The photographs have been received.

This picture doesn't leave any question of doubt in my mind that the big goose really exists. I have been trying for years to convince Wetmore and some of the other scientific men of this fact, and I think your specimen is going to turn the trick. Am going to Washington tonight and talk to Dr. [E. W.] Nelson about it.

In a letter of July 14, 1922, to E. W. Nelson, then Chief of the United States Bureau of Biological Surveys, Mershon enclosed a letter from the secretary of the North Dakota Game and Fish Board stating that five large geese were on their game farm. In his reply to Mershon on July 17, 1922, Nelson stated that he was sending one of the Bureau's best experts, Dr. Wetmore, to photograph and make other observations concerning these geese.

I had in mind if they proved to be an undescribed form to describe them when I secure the material to be obtained by Dr. Wetmore. The matter is a most interesting one and I am inclined to think the goose really represents an undescribed geographic race peculiar to that region.

A. K. Fisher, Acting Chief of the Bureau, to W. B. Mershon, December 5, 1922:

With regard to your letter of Dec. 2 we are pleased to report that Mr. L. C. Pettibone, of Dawson, North Dakota, under date of Nov. 25, 1922, forwarded to us one of the large geese concerning which there has been so much interest recently. The bird in question, which came through in fine shape, thanks to Mr. Pettibone's care in packing, is a gander fully adult, that weighed 14½ lbs. The bird was exceedingly fat and in measurements does not surpass others of lesser weight that we have in the National Museum. *It exhibits the pale, grayish coloration ascribed to the native goose of the region in question.* [Italics by present author.]

R. P. Holland to W. B. Mershon, July 13, 1922:

I have killed a great many geese, but it was only about six years ago that I began weighing them and trying to get one of the big boys as a specimen.

What interested me in this was the statements of gunners from along the Platt River and north of this section, although "Widgeon" [1922] states that he saw these birds along the Arkansas river near McPherson, Kansas. I believe the Platt River was about the southern limit of their range.

I am like you and I really don't care what they name it just so they admit such a bird really did exist, and probably still is to be found in the Dakotas. As I told you in my former letter, most ornithologists do not admit that the ordinary gunner knows anything about birds.

John C. Phillips to W. B. Mershon, January 15, 1923:

A few days ago I was in Washington and saw the two live Dakota geese which Wetmore brought back from the West. They looked large *and had a curious white ring around the neck below the black of the neck. . . .* Wetmore thinks there is doubt of a race of large geese there in Dakota, but undoubtedly they intergrade as far as size goes with geese from other places. [Italics by present author.]

George Bird Grinnell to W. B. *Mershon, July 18, 1922:*

I heartily congratulate you on having discovered some of these big geese in the flesh. They should be guarded by the people who have their custody as the apple of one's eye.

I should hesitate a little bit about killing them at the present time,— certainly not until you hear from Phillips. The taxidermists at Mandan, and the one at Grafton, North Dakota, ought to be communicated with and urged to make skins of any of these big geese that they can get hold of, that may be killed near them. I would not kill these living birds at present. If you wait a while — meantime getting Mr. [G. M.] Hogue more enthusiastic on the subject, these geese may mate and we may have a supply of them. You do not say how old these birds are, but I conclude that they are young — possibly two years old. This is a matter of very great interest; and I hope, and of course believe, that you will handle it in a judicious way. One thing ought to be done, I feel. The living birds should be photographed, if possible, in different lights, though perhaps it is doubtful whether any great results would come from that.

Joseph Grinnell to W. B. *Mershon, May 31, 1923:*

With regard to the big goose in North Dakota, it seems to me as though you would be justified in expending a whole lot of energy and substance in determining the facts, *before it is too late.*

If you could put in a season there, or have some one else do so, you would insure the recording of the facts for all time. There is a good deal of natural history that is going to be lost to science altogether, if people who realize this don't get busy right away. There are cases in point right here in California that I am personally working on. Scientists of the future will look back and blame *us* for neglecting *our* opportunities, just as we now deplore the inaction of people one hundred years ago who, seemingly, had chances to record facts about the birds then common, but now scarce or gone.

E. W. *Nelson to* W. B. *Mershon, March 15, 1924:*

I have read the carbon copy of your letter to Mr. R. P. Holland regarding the large goose of the Dakotas with great interest, and note what you say regarding the Biological Survey in that communication.

In the fall of 1922 when Dr. A. Wetmore of this Bureau was in North Dakota endeavoring to secure specimens of these large geese, he met Mr. [J. J.] Gokey while at Dawson and had an interesting talk with him regarding geese and hunting in general. Dr. Wetmore was told of the nesting site that Mr. Gokey mentioned to you, but found that the birds had been driven out by increase in settlement in that region. It is possible that a pair or two still breed in Kidder County, but it is exceedingly difficult to get news of them and to secure specimens. You will probably recall that Mr. Lee Pettibone sent us one of these birds that tipped the scales at 14½ pounds, so that we have this specimen. In addition we secured a living pair from Mr. Elmer Judd, which we have had here in the National Zoological Park, and had hoped that they might breed. Unfortunately, one of them died, or was killed in some way, last summer, so that there is only the single bird remaining.

. . . .

If these large geese should eventually prove separable as a distinct subspecies inhabiting a section of the northern plains country, there is already a name available that may be applied to them, so that it will not be necessary to describe the bird as actually new or unknown.

W. C. Henderson to W. B. Mershon, April 28, 1924:

If this bird is really distinct from other Canada geese, it is at most a subspecies according to the present day usage of ornithologists in such matters. The name that may be applied to this third, in case one is really necessary, is still a matter for speculation.

We have been very glad to have your comments on this matter, and shall send you the full account of all findings in regard to this bird when published.

E. W. Nelson to W. B. Mershon, August 5, 1924:

Now about the big, pale-colored Canada goose from North Dakota and adjacent part of Canada. I am inclined to believe that there is a large, pale-colored subspecies of Canada geese in that region which has been shot out until it is now rather sparsely represented. This conclusion was the result of an investigation made by Doctor Wetmore, but unfortunately we were not able to secure a sufficient number of specimens to make the determination absolutely positive. [This was an entirely defensible decision; too often in the past races had been described on the basis of an inadequate number of specimens.] So far as we have gone, however, I think you are justified in believing that this goose represents a recognizable subspecies.

This large pale-colored goose, however, appears to have been given a scientific name long ago, so that when the matter is absolutely determined through the collection of a sufficient number of specimens to establish it beyond doubt it will result in the revival of that name and the recognition of that subspecies. The delay incident to publishing on the subject is due merely to our desire to secure sufficient material to make a good statement of the case. The weights you give of the geese taken undoubtedly are of representatives of this bird. Should you at any time get additional specimens of it, I shall be very glad indeed to have you send them to the Biological Survey; or if you send them to any other institution, please notify me in order that we can borrow them for study. Any additional facts you secure I shall greatly appreciate.

I regret to say that apparently this goose is going the way of the whooping crane. It appears to have occupied about the same breeding range. The occupation of their breeding territory through the development of farming is no doubt mainly accountable for its decrease.

E. W. Nelson to W. B. Mershon, April 30, 1925:

From the information you give in your article and the mounted specimen you have I am strongly inclined to believe that a large, pale subspecies of the Canada goose *formerly bred in the northern Great Plains region, and thence into the prairie country of Southern Canada.* [Italics by present author.]

E. W. Nelson to W. B. Mershon, December 23, 1925:

I am glad to have the additional information in regard to the gray goose and hope in the not distant future, as indicated in my letter of April 30, to prepare an article covering results on our investigations concerning this form.

R. P. Holland to W. B. Mershon, January 8, 1931:

I think we convinced the Biological Survey, but John Phillips cooked our goose for us.

H. L. Betten, Alameda, California, to R. P. Holland, June 18, 1934:

Am enclosing a letter from my old friend, Feudner, which please return; it is gratifying to me on two counts: first, because Otto sustains my contention as to big geese; secondly, because he refers to big ones still living, including some in captivity. Of course, I have known absolutely large honkers had existed and ever since . . . cocksure ornithologists ridiculed the idea I have aimed to nail their hides to the barn door. One of our locals, who is just past the diaper stage, sarcastically stated: "These old oracles merely estimate the weight of mythical big geese and their imagination expands with the years. I have never seen . . ." etc. etc. and etc. (There is a hell of a lot he hasn't seen.)

If possible I will arrange a trip to the old Feudner farm with Otto and Geo. Tonkin of Biological Survey, weigh some of those live geese on Fairbanks scales, take pictures and secure the necessary affidavits — after which a tale about big geese.

W. B. Mershon to R. P. Holland, October 29, 1934:

I have a whole letter file full of this subject that I will be very glad to send to you for permanent filing, if it would be of any use. I am too old to use it any more, but it contains original letters from dozens of correspondents who give the weights and give information about those big geese, which both you and I believe was entirely separate and distinct from the old honker. I have in my collection a mounted gosling that weighed over 14 lbs. which was killed in Kidder County, N.D.

W. B. Mershon to R. P. Holland, November 12, 1934:

Wetmore went to Dawson, N.D. purposely to look into this goose business and my recollection is he wrote me they had a specimen and that the goose had already been named. However, why don't you take it up with him again? See if you can get anything definite. There has always been something a little fishy about this whole transaction as far as Washington is concerned.

A. Wetmore to R. P. Holland, November 27, 1934:

Since our earlier correspondence on this matter, there is a young chap in California, named James Moffitt, whose address is 1879 Broadway, San Francisco, California, who has taken up a thorough study of all of the American geese with the idea of trying to straighten out many problems that concern them. Mr. Moffitt has been here in the East for the past few weeks visiting large collections and examining specimens. I think that he has more information on the subject than any one else has previously acquired and bids fair to do a good job. I talked with him several times when he was here in Washington and we discussed tentatively the large goose that you mention but did not go into it in detail.

R. P. Holland to Dr. Alexander Wetmore, November 28, 1934:

I have a letter from John Phillips, in which he says, "I believe you are right in your surmise that there was a larger race of Canadians in the Dakotas, Nebraska and southern prairie region, but unfortunately, nothing tangible remains on which to base a new subspecies."

A. Wetmore to R. P. Holland, January 15, 1935:

Young Moffitt, concerning whom I wrote you, will, I think, get up successfully the matter of the large geese of the interior. He has seen most of the specimens available in museums in this country at the present time, and is working enthusiastically over his data to the exclusion of any other ornithological interest. I have a long letter from him on my desk with regard to his recent trip east to which I have not yet made reply. He examined one large bird [an immature male] that Lee Pettibone sent me from North Dakota in the fall of 1922.

R. P. Holland to W. B. Mershon, March 4, 1935:

Have yours of the 25th, and also copy of the Nebraska Bird Review, with your note on the big goose. Where in the hell do you get this Holland goose stuff. It will either be Mershon-Holland goose or just Mershon goose. You have done a lot more work in digging up this stuff than I have. All I can do is keep prodding these birds down in Washington with the hope of getting some action.

W. B. Mershon to R. P. Holland, March 9, 1935:

You ask where I get this Holland goose stuff. You started the ball rolling and it is your stuff if they ever do come to name it. It is all right to make it the Holland-Mershon investigation but I probably never would have done anything about it if you had not taken the initiative.

R. P. Holland to H. M. Laing, Comox, B.C., January 25, 1937:

I am afraid the Holland-Mershon goose will never take its place in the scientific world, but I firmly believe such a species existed. It was not a brown goose. He was a little grayer than the Canada. He rarely, if ever, traveled with smaller geese. Flocks of six or eight seemed to live together. If a gunner got into a flock of these birds they were all big ones.

W. B. Mershon to R. P. Holland, October 26, 1937:

My health has not improved any; in fact, I think I am just a little nearer the final hitching post than I was six months ago. I will be eighty-two in January and have long since done my life's work.

W. B. Mershon to H. L. Betten, March 17, 1939:

Now, as I said before, I am going to send Ray Holland a copy of this letter and you two will have to continue this stirring up as to what the proper name should be of this big goose whose weight exceeds that of the Canada. I killed three of them once out of a flock of five and all weighed 14 lbs. and over.

R. P. Holland to W. B. Mershon, March 23, 1939:

Got the copy of your letter to H. L. Betten. I will go the limit on the big goose thing, but we went up against a stone wall hard when John Phillips said definitely that your big goose was simply Branta canadensis. Poor John has checked out by now; but when they get into the records at Washington, they will find all his comments, and I simply could never raise the least spark of recognition from him.

W. B. Mershon to R. P. Holland, March 30, 1939:

It is going on four years since I have been able to walk any distance, and for nearly a year I have not been able to take a step even with crutches.

When you get around to it, you stick to that big goose subject. John Phillips, dear fellow, was certainly mistaken.

I am longing for green grass to come. The grackle and robins are here from the south but that is about all.

Ira N. Gabrielson to R. P. Holland, April 3, 1939:

A competent ornithologist, not a member of the Biological Survey or National Museum staff, is now engaged in making a thorough review of the entire question as to the classification of the geese. He has examined the various forms and worked out the diagnostic characters that distinguish the various races. It is hoped that he will be able to publish this material in a reasonable time. The information that we have regarding his findings to date is confidential in character. We are hoping that the results of his studies will be published in the near future so that we can use them in our work and be free to tell the world what we know about the geese.

W. B. Mershon to R. P. Holland, April 13, 1939:

These birds bred in Kidder County. They were a different color, different shape [Fig. 2], kept by themselves and fed separately from the big flight of Canadas, Hutchins, and other geese that were there in the late 80's and early 90's in tens of thousands.

I am dictating this from bed, flat on my back. Thermometer only 20° this morning. I do hope when warm weather does come — if it ever does — that I will be able to be wheeled out in my chair and sniff the spring air again.

William B. Mershon died in 1939 at the age of 84, his dream of scientific recognition for the giant Canada goose not realized during his lifetime.

R. P. Holland to H. L. Betten, January 19, 1940:

I wish we could get the big goose matter straightened out. Dr. Nelson admitted to me that he was convinced that I was right in my claims, but that without a specimen science couldn't act. Then Bill Mershon got what he said was a specimen and Dr. John Phillips turned it down definitely and classed it as nothing more than an extra large *Branta canadensis* [*canadensis*], overstuffed by a poor taxidermist.

In 1944, McAtee, in a short note in the *Auk* entitled "Popular Subspecies," remarked that "the northern plains goose seems even to have the geographic qualifications for a subspecies . . ." and urged ornithologists to investigate the possibility.

Aldrich (1946:97) described the race *B. c. moffitti* and included in its range most of the range of *maxima*. It is therefore not surprising that Conover (*in* Hellmayr and Conover, 1948:303–4) wrote regarding the race *moffitti*: "This is a rather unsatisfactory race, the distinguishing characters standing out fairly clearly only in the specimens from the most westerly part of its range as given."

Recognition of the giant Canada goose as a separate race was contained in the notes left by James Moffitt who was killed in World War II. In a publication released on November 12, 1951, Jean Delacour, basing his studies on Moffitt's extensive notes, described the giant Canada goose, *Branta canadensis maxima*.

In mid-January, 1962, I studied the flock of Canada geese wintering at Rochester, Minnesota. Measurements taken during banding operations and the collection of nine skins constituted the basis for the rediscovery of the giant Canada goose.

Ray P. Holland, now living in New Mexico, succinctly summed up Mershon's emotional involvement with the giant Canada goose in a letter to me November 12, 1962:

> Capt. Billy Mershon was a great fellow. . . . His big interest was to get the big goose named the Mershon-Holland goose. . . . He would much rather have the big goose named for him than to have struck another vein of copper.

The mills of science, conservation, and officialdom often grind slowly and finely — and sometimes too late. It may be well to ask why has adequate study of another extremely rare goose, the Tule goose (*Anser albifrons gambeli*), a giant race of the white-fronted goose, lagged for decades when it has long been apparent that special study and protective measures were necessary to insure its survival?

Physical Characteristics 2

MOST of the races of Canada geese are readily distinguished, but as a group they exhibit a series of clines or gradations in weight, size, and body proportions.[1] It is sufficient to state at this point that the giant Canada goose (*B. c. maxima*), being the largest of the races, exhibits the extremes of many of these evolutionary trends. Next to it in size are *B. c. moffitti*, the western Canada goose; *B. c. interior*, the Hudson Bay Canada goose; and *B. c. canadensis*, the Atlantic Canada goose. The Queen Charlotte goose, *B. c. fulva*, sometimes equals the above three races in weight but is usually smaller in size. For the convenience of the reader, the distribution and chief characteristics of the other three large races of Canada geese are briefly summarized before discussing the characteristics of the giant Canada goose and making comparisons between it and the following races (*see also* note 1, Chapter 15, "Discussion"):

Branta canadenis canadenis (LINNAEUS)

Atlantic Canada goose (*fig.* 3)

DISTINGUISHING CHARACTERS — A light-colored race with a whitish mantle over the forepart of the back; underparts grade from medium gray on the breast to almost white at the base of the neck.

BREEDING RANGE — Nova Scotia, Anticosti Island, Newfoundland, and in Newfoundland-Labrador from the height of land north of the Gulf of St. Lawrence northward to the tree line at Webb's Bay on the Labrador coast (Austin, 1932), and westward in Ungava to the height of land where it gradually intergrades with the race *B. c. interior*.

WINTERING RANGE — Vicinity of Port Joli and Port Hebért, Nova Scotia, and along the Atlantic coast from southeastern Massachusetts south to Pea Island, North Carolina.

Branta canadensis interior TODD

Hudson Bay Canada goose (*fig. 4*)

DISTINGUISHING CHARACTERS — A dark race with the breast ranging from a medium gray to slate gray; the back is dark brown and lacks a conspicuous light mantle over the forepart.

BREEDING RANGE — Hudson Bay Lowlands of Manitoba, Ontario, and Quebec, and in the latter province extending northward to Hudson Straits and eastward to the height of land where it intergrades with *B. c. canadensis;* south of the Fort George River it is chiefly confined to the low coastal belt of James Bay; breeds on the islands of Hudson and James bays, notably on the Belcher, Twin, Akimiski, and Charlton islands and on many of the smaller coastal islands along the east coast from the Nastapoka Islands southward.

WINTERING RANGE — Southeastern South Dakota (small numbers) and Swan Lake, Missouri, southward to the Louisiana Gulf Coast and eastward to Lake Mattamuskeet, North Carolina, but not including the populations wintering on the Tennessee River and in the general region of St. Marks, Florida, which represent an undescribed race (Fig. 4).

Branta canadensis moffitti ALDRICH

Western Canada goose (*fig. 5*)

DISTINGUISHING CHARACTERS — A light-colored race; lacks the whitish mantle over the back present in *canadensis.*

BREEDING RANGE — Central and southeastern British Columbia, eastern Washington, Oregon, northeastern California, northern half of Nevada and Utah, and eastward in Montana and Wyoming to the Continental Divide.

WINTERING RANGE — Chiefly in central California, west-central Nevada, southern California (Imperial Valley), and the area adjacent to the northern portion of the Gulf of California.

In most subsequent discussions, these races will be referred to by their subspecific name only.

To distinguish geese in their second year of life (yearlings) from older adults, it is necessary to use cloacal characters (*see* Chapter 8, "Characters of Age and Sex and Sexual Maturity"). For this reason, data on museum skins of these age classes are grouped as adults. The sex and age of the live geese handled were determined by the criteria described in the chapter on age and sex determination. During the fall and winter period, three age-classes were distinguished: *immatures*, 5 to 8 months old; *yearlings*, 17 to 20 months; and *adults*, 29 or more months of age. Museum skins frequently lacked complete age and sex data or, in a number of cases, it was apparent from plumage characters or the possession of a well-developed "wing spur" that the specimen had been incorrectly aged and sexed at the time of collection. Only data from specimens whose age and sex could be definitely established were used.

Except for measurements of live geese taken at the Waubay National Wildlife Refuge, South Dakota, all measurements presented in the tables were made by me. Measurements given for wing length are of the straightened and flattened wing as opposed to wing chord.

WEIGHT

The early fame of the giant Canada goose rested almost entirely on its weight, particularly that of exceptionally large males (Figs. 6 and 7), which received attention far out of proportion to the mode of the population. The extraordinary weights attained by some individuals were first documented by Mershon (1925):

> Now as to the weights of these big fellows, which we all recognized as a different goose from the ordinary Canada. When in North Dakota, some time in the late eighties, '87 or '88, a gentleman in a private car adjoining our old "City of Saginaw" hunting car came in with a Goose that he had weighed and claimed that it weighed 18 lb. To my mind that stood as a record, and we were trying to obtain an equally large or larger one, so that we weighed our Geese.
>
> We frequently had them 14 and 15 lb. My friend Sanford Keeler, at that time General Superintendent of the Pere Marquette Railroad here in Saginaw, accompanied me north of Dawson one day and we shot on the north side of Lake Sibley. We got a number of large Geese. He killed one that weighed 17¼ lb., and I shot one that weighed 17 lb.

In his article, Mershon (1925) listed many of the weights of large geese that came to his attention. These records, additional ones from the Holland-Mershon files, and a number compiled by me are given in Table 1. They have value beyond their novelty interest. To those familiar

1. Records of giant Canada geese that weighed 16 pounds or more

Locality	Weight (Pounds)	Year	Authority
ALBERTA			
Edmonton (?), Wainwright (?)	16¾[1], 16, 16	1940's (?)	W. F. H. Mason[3] — G. H. Nicholson. *Rod and Gun* (clipping, year unknown)
Inglewood Refuge, East Galgary	15, 16		
SASKATCHEWAN			
Johnstone Lake	16	1921	William B. Mershon, Jr.[3]
Foam Lake	16+	1961	Harvey W. Miller[5]
Alsask	18	1930	Gene McBride. *Rod and Gun.* 32(6):420
MANITOBA			
North Shoal Lake	22	1943	Anon. 1945 and *Winnipeg Tribune*
Mariapolis	17½	1953	The *Dufferin Leader.* November 19, 1953
Whitewater Lake	16½	1912	Angus Shortt
Gladstone	18	1950	Eugene Bossenmaier[5]
BRITISH COLUMBIA			
Unknown	17	unknown	R. C. Mayne (1862)
Northern B. C.	18	unknown	J. Petley[5]
OREGON			
Harney Co.	17¼	1940	Cecil S. Williams[5]
Harney Co.	16	1940	Cecil S. Williams[5]
CALIFORNIA			
Unknown	19	1923	H. L. Betten[3]
Unknown	17, 22	1929, 1930	H. L. Betten[3]
UTAH			
Otter Creek Reservoir	16	1939	Cecil S. Williams[5]
COLORADO			
Unknown	51¼ (total weight of 3 geese)	1889	George M. Sibley[3]
NORTH DAKOTA			
Jamestown, Stutsman Co.	18, 20, 20	1962	Carl E. Strutz[2, 5]
Rogers, Barnes Co.	18	1887	A. P. Paulson[3]
Rogers, Barnes Co.	21	1886	A. P. Paulson[3]
Unknown	23	prior to 1922	George M. Hogue[8]

1. Minus head and neck. 2. Captive. 3. Mershon and Holland *letters.* 4. *Winnipeg Tribune.* 5. *Personal Communication.* 6. By Glenn Smart and

Locality	Weight (Pounds)	Year	Authority
Lake Irvine, Ramsey Co.	23	Late 1870's	Charles C. Wenz[3]
Oakes, Dickey Co.	$18\frac{3}{4}$	1947	Kenneth Brossmann[5]
Pierce Co.	19	unknown	M. O. Steen[3]
Kidder Co.	17	unknown	J. J. Gokey[3]
Steele Co.	20, 23	1909	George M. Hogue[3]
Carrington, Oster Co.	17, 18	Early 1890's	Charles E. Deane[3]
Dawson, Kidder Co.	$16\frac{1}{4}$ 17, $17\frac{3}{4}$, 18	1886	J. J. Gokey[3]
Dawson, Kidder Co.	19	1886	L. C. Pettibone[3]
SOUTH DAKOTA			
Elk Point, Union Co.	19	1906	James Chausser[3]
Elk Point, Union Co.	22	ca. 1906	James Chausser[3]
TEXAS			
Olney (?), Young Co. (?)	18	unknown	W. P. Reynolds[3]
MINNESOTA			
Kandiyohi Co.	$19\frac{1}{2}$	1895	A. C. Buethe[5]
Crookston, Polk Co.	22	ca. 1900	Harry Jensen[5]
Heron Lake, Jackson Co.	24	ca. 1900	Morton Barrows[3]
Round Lake Waterfowl Station, Jackson Co.	$17\frac{5}{16}$[2]	1962	This paper[6]
Waseca, Waseca Co.	18	1918	Arnold Wolfers[5]
Rochester, Olmsted Co.	16	1953 or 54	Byron Brueske[5]
IOWA			
Riverton, Fremont Co.	18	1961	Jack Musgrove[5]
Emmettsburg, Palo Alto Co.	$16\frac{1}{2}$[2]	unknown	A. J. Poeppe[5]
MISSOURI			
Platte Co.	21	1915	R. A. Brown[7]
WISCONSIN			
Rock Co.	$19\frac{1}{2}$	1924	A. J. Rusch, Wisc. Con. Dept.
Horicon Marsh, Dodge Co.	16	1964	Richard A. Hunt
ILLINOIS			
Horseshoe Lake, Alexander Co.	16, 17, 19	1941	William Brown[5]
Southern Illinois	16	unknown	F. Henry Yorke. *Amer. Field* 35(9):218

author. 7. *Notarized letter*, May 16, 1963 to Harold H. Burgess. 8. *New York Times*, November 6, 1922; *Madison* (Wis.) *State Journal*, December 26, 1922.

with other large races of Canada geese, some of these past weight records may appear to be questionable; however, these data are fully matched by living specimens and by giant Canada geese shot in recent years (Figs. 8, 9, and 10).

Weights recorded for various flocks of *maxima*, *interior*, and *moffitti* are summarized in Tables 2 and 3. Differences in average weights between populations of *maxima* and *interior* are given in Table 4. A considerable disparity between the weights of various populations of *maxima* will also be noted; however, the average weights for all populations of *maxima* are higher than those for the race *interior* which has been more intensively studied in this respect than any other population of waterfowl (Hanson, 1962*b*).

The Round Lake *maxima* averaged markedly heavier than the Manitoba populations (from which the Rochester flock is derived). Included in the averages for the Round Lake population are both pinioned and flying birds. No important differences in weight between the two groups were noted. Weights of a captive Illinois population of immature *maxima* examined in the fall of 1962 were similar to the Round Lake population, but in respect to length of middle toe they exceeded the Round Lake geese. Giant Canada geese of very large size were also observed on the farms of Oscar Luedtke, Lotts Creek, Kossuth County (Fig. 11), and A. J. Poeppe, Emmettsburg, Palo Alto County, Iowa; Emmett Kern, Granada, Martin County, Minnesota; Carl E. Strutz, Jamestown, Stutsman County, North Dakota; and at Denver, Colorado. The origins of these populations are discussed in Chapter 3, "Breeding Range," but all these flocks are descendants of former game farm and decoy stocks.

The extent to which the weight and size of the geese in these flocks reflect a long history of captivity and selection for size cannot be assessed completely with data now available. But it is highly probable that, except for the possibility of their having large amounts of body fat, most present stocks in the United States that are of captive origin do not differ from wild stocks which once nested in their respective areas. Two of the present-day populations, which in respect to phenotype and genotype are believed to be most similar to the original regional stocks, are the flock of free-flying giant Canada geese now resident most of the year on the Waubay National Wildlife Refuge, Day County, South Dakota, and the captive flock of Carl E. Strutz at Jamestown, North Dakota. Both flocks are derived from early regional wild stocks; the Waubay flock is descendent from a local captive flock which, in turn, was propagated from eggs of wild Canada geese collected in Day County.

According to Robert R. Johnson, manager of the Waubay refuge, Canada geese weighing 18 pounds have been reported shot in Day County nearly every year, but the authenticity of these records had always been questioned. In the fall of 1962, at my instigation, Johnson interviewed a number of local hunters regarding weights of large geese they had shot. The following records were obtained: two geese, each weighing 12½ lb.; two, each weighing 13½ lb.; and one, weighing 16½ lb.

Other miscellaneous weight records for *maxima* contributed by various investigators might best be summarized here for comparison. At Sand Lake National Wildlife Refuge the following weights were obtained of geese shot in late November to early December, 1961: adult males, 11 lb. 6 oz and 12 lb. 10 oz; immature male, 10 lb. 8 oz; adult female, 11 lb. 4 oz (Lyle Schoonover, *personal communication*). Noted on recovery records of two geese banded at Rennie, Manitoba, and shot in Rock County, Wisconsin, were the weights 13 lb. 3 oz and 12 lb. 3 oz (Richard A. Hunt, *letter*, January 9, 1963).

Several small samples of weights of *maxima* shot in Manitoba were obtained in the fall of 1962: from 50 miles northwest of Portage la Prairie, adult female, 10 lb. 11 oz; immature female, 9 lb. 14 oz (Robert McWhorter, *personal communication*); adult female, 10 lb. 11 oz; and immature female, 9 lb. 14 oz (possibly from Saskatchewan, Eugene F. Bossenmaier, *personal communication*, December 11, 1962). From The Pas region, the following weights were obtained on September 28 and 29, 1962 (Jack L. Howard, *letter*, January 14, 1963): immatures, 8.9 lb., 9.4 lb., 10.4 lb., and 11.5 lb.; adults, 9.7 lb., 9.9 lb., 10.3 lb., 11.0 lb., 11.4 lb., 12.2 lb., 12.4 lb., 12.6 lb., and 13.6 lb. All but the first three weights in the adult series are well above weights of the great majority of adult males at Horseshoe Lake, Illinois (Table 2).

There are authentic records of giant Canada geese weighing 19 pounds or more (Table 1), although it must be admitted that in the early stages of the study I regarded such records as fictional. For example (Anon., 1945a:10):

> Second only to the Swan in size, the common Canada goose has been known to reach 22 pounds in weight, a bird of that size having been taken at North Shoal Lake, Manitoba, in 1943.

The above goose and two other unusually large geese killed on October 15, 1943, were reported to have been shot from a flock of about

2. Body weights in grams of *Branta canadensis maxima*, *B. c. interior* and *B. c. moffitti*

Subspecies by age-sex class	Locality
IMMATURE MALES	
B. c. maxima	Round Lake, Minn.
B. c. maxima	Southern Manitoba
B. c. maxima	Rochester, Minn.
B. c. interior	Horseshoe Lake, Ill.
YEARLING MALES	
B. c. maxima	Rochester, Minn.
B. c. interior	Horseshoe Lake, Ill.
B. c. moffiti	Brigham City, Utah
ADULT MALES	
B. c. maxima	Round Lake, Minn.
B. c. maxima	Southern Manitoba
B. c. maxima	Rochester, Minn.
B. c. maxima	Mosquito Creek, Ohio
B. c. interior	Akimiski I., N.W.T.
B. c. interior	Horseshoe Lake, Ill.
B. c. moffiti	Brigham City and Richfield, Utah
IMMATURE FEMALES	
B. c. maxima	Round Lake, Minn.
B. c. maxima	Southern Manitoba
B. c. maxima	Rochester, Minn.
B. c. interior	Horseshoe Lake, Ill.
B. c. moffiti	Richfield, Utah
YEARLING FEMALES	
B. c. maxima	Rochester, Minn.
B. c. interior	Horseshoe Lake, Ill.
ADULT FEMALES	
B. c. maxima	Round Lake, Minn.
B. c. maxima	Southern Manitoba
B. c. maxima	Rochester, Minn.
B. c. maxima	Mosquito Creek, Ohio
B. c. interior	Akimiski I., N.W.T.
B. c. interior	Horseshoe Lake, Ill.

25 (*Winnipeg Free Press*, November 5, 1943). It is evident from the photograph which accompanied this article that this goose was "gigantic."

Records of extreme weights and wing spans are of interest in themselves, but such data are also useful in deducing the range and migration

Date	Seasonal Activity	Number	Average	Range	Standard deviation
Dec. 6–8	Wintering	9	5,963 ± 249	5,040–7,569	747
September	Late body molt	25	3,970 ± 17	3,370–4,700	86
	(flying)				
Jan. 29–30	Wintering	20	4,261 ± 94	3,430–5,075	418
January	Wintering	37	3,615 ± 52	—	318
Jan. 29–30	Wintering	11	4,411 ± 64	4,111–4,763	212
January	Wintering	7	3,960 ± 94	—	249
Hunting season	—	1	4,275	—	—
Dec. 6–8	Wintering	7	6,525 ± 337	4,940–7,484	893
Sept. 2–9	Late body molt	9	4,851 ± 100	4,220–5,270	299
	(flying)				
Jan. 29–30	Wintering	13	4,884 ± 98	4,196–5,415	354
Jan. 8	Wintering	8	6,132 ± 207	5,216–6,804	587
Aug. 10–14	Molting	18	4,349 ± 118	2,890–4,870	500
	(flying)				
Nov. 1–Dec. 31	Wintering	31	4,069 ± 55	—	304
Hunting season	—	3	4,093 ± 68	3,990–4,220	290
Dec. 6–8	Wintering	3	5,245 ± 401	4,760–6,040	694
Sept. 2–9	Late body molt	16	3,487 ± 32	3,180–4,140	130
	(flying)				
Jan. 29–30	Wintering	15	3,821 ± 81	3,430–4,337	315
January	Wintering	35	3,071 ± 41	—	249
Hunting season	—	1	3,080	—	—
Jan. 29–30	Wintering	11	3,690 ± 74	3,430–4,220	247
January	Wintering	4	3,466 ± 157	—	313
Dec. 6–8	Wintering	13	5,514 ± 166	4,270–6,435	598
Sept. 2–9	Late body molt	8	3,871 ± 70	3,600–4,070	197
	(flying)				
Jan. 29–30	Wintering	7	3,868 ± 101	3,572–4,167	267
Jan. 8	Wintering	5	5,387 ± 377	4,536–6,350	845
Aug. 10–14	Molting	12	3,786 ± 75	3,350–4,180	259
	(flying)				268
January	Wintering	10	3,561 ± 85	—	

routes of this subspecies; therefore, additional information of this kind is presented in later discussions.

How can the occurrence of adult male geese of extreme size and weight be explained biologically? The following relationships provide a reasonably complete explanation: (1) There is evidence of a general

3. Body weights in grams of *Branta canadensis maxima* and *B. c. interior* during the incubation and flightless period of the molt

Subspecies by age-sex class	Locality
YEARLING MALES	
B. c. maxima	Thelon River, N.W.T.
B. c. maxima	Missouri: Trimble Wildlife Area
B. c. maxima	Missouri: Trimble Wildlife Area
B. c. interior	Akimiski I., N.W.T.
ADULT MALES	
B. c. maxima	Manitoba: Dog and Chitek lakes
B. c. maxima	Saskatchewan: Kinistino
B. c. maxima	Alberta: Farrell Lake
B. c. maxima	South Dakota: Waubay, N.W.R.
B. c. maxima	South Dakota: Sand Lake, N.W.R.
B. c. maxima	Missouri: Trimble Wildlife Area
B. c. maxima	Missouri: Trimble Wildlife Area
B. c. interior	Akimiski I., N.W.T.
YEARLING FEMALES	
B. c. maxima	Thelon River, N.W.T.
B. c. maxima	Missouri: Trimble Wildlife Area
B. c. maxima	Missouri: Trimble Wildlife Area
B. c. interior	Akimiski I., N.W.T.
ADULT FEMALES	
B. c. maxima	Manitoba: Dog and Chitek lakes
B. c. maxima	Saskatchewan: Kinistino
B. c. maxima	Alberta: Farrell Lake
B. c. maxima	South Dakota: Waubay N.W.R.
B. c. maxima	South Dakota: Sand Lake N.W.R.
B. c. maxima	Missouri: Trimble Wildlife Area
B. c. maxima	Missouri: Trimble Wildlife Area
B. c. interior	Ontario: Sutton River
B. c. interior	Akimiski I., N.W.T.

Date	Period or activity	Number	Average	Range	Authority
July 2, 1964	Molting	1	4540	—	E. Kuyt (ltr., Sept. 8, 1964)
June 17, 1963	Molting	13	4593	—	G. Brakhage (ltr., Nov. 21, 1963)
July 2, 1963	Molting	18	4308	—	G. Brakhage (ltr., Dec. 18, 1963)
July 16–21, 1958, and 1959	Molting	18	3853	3425–4065	Hanson (1962a)
July 19, 1962	Molting	3	4477	4200–4800	This paper
May 29, 1962	With 1-week-old brood	1	4780	—	Alex Dzubin (ltr., Jan. 31, 1963)
May 5, 1962	Incubating	1	4610	—	Alex Dzubin (ltr., Jan. 31, 1963)
July 9, 1963	Molting	13	4104	3685–4905	Ray D. Hart (unpublished data)
July 10, 1963	Molting	7	4192	3686–5018	Lyle Schoonover
June 17, 1963	Molting	47	4886	—	G. Brakhage (ltr., Nov. 21, 1963)
July 2, 1963	Molting	55	4626	—	G. Brakhage (ltr., Dec. 18, 1963)
July 10–Aug. 7, 1958 and 1959	Molting	45	3946	3140–5135	Hanson (1962a)
July 4, 1964	Molting	1	3859	—	E. Kuyt (ltr., Sept. 8, 1964)
June 17, 1963	Molting	18	4026	—	G. Brakhage (ltr., Nov. 21, 1963)
July 2, 1963	Molting	21	3742	—	G. Brakhage (ltr., Dec. 18, 1963)
July 10–27, 1958 and 1959	Molting	20	3235	2965–3545	Hanson (1962a)
July 19, 1962	Molting	2	3858	3720–3995	This paper
May 29, 1962	Incubating, and with 1-week-old brood	2	3200	3125–3275	Alex Dzubin (ltr., Jan. 3, 1963)
May 5, 1962	Incubating	1	3790	—	Alex Dzubin (ltr., Jan. 3, 1963)
July 9, 1963	Molting	9	3453	3289–3799	Ray D. Hart (unpublished data)
July 10, 1963	Molting	8	3721	3402–4082	Lyle Schoonover
June 17, 1963	Molting	61	4193	—	G. Brakhage (ltr., Nov. 21, 1963)
July 2, 1963	Molting	74	3830	—	G. Brakhage (ltr., Nov. 21, 1963)
May 28–June 6, 1959	Incubating	8	3287	2925–3840	Hanson (1962a)
July 12–Aug. 8, 1958 and 1959	Molting	30	3349	2815–3870	Hanson (1962a)

4. Differences by per cent in body weight and wing length between populations of *Branta canadensis maxima* and between *B. c. maxima* and *B. c. interior*

Populations compared	Difference in body weight (per cent larger)		Difference in wing length (per cent larger)	
	Males	Females	Males	Females
Round Lake maxima to S. Manitoba maxima	34.5	42.4	2.2	1.1
Round Lake maxima to Rochester maxima	33.6	42.6	4.3	2.2
Round Lake maxima to Horseshoe Lake interior	60.4	54.8	5.9	4.5
S. Manitoba maxima to Rochester maxima	—	—	2.0	1.0
S. Manitoba maxima to Horseshoe Lake interior	—	—	3.6	3.3
Rochester maxima to Horseshoe Lake interior	20.0	8.6	1.5	2.3

south to north clinal decrease in the average size of individuals in various populations of *maxima*. (2) Geese of exceptional weight represent notable extremes within a normal distribution. (3) Very heavy individuals are a distinct physical type (Fig. 12), differing from others of average weight by their greater stockiness and muscular development. Such individuals can be recognized by their thick necks; in contrast, other males in the same population may have more slender, snake-like necks (Fig. 13). These differences in body build have a genetic basis, judging from information I have gained in discussions with several owners of captive flocks. I have noted similar variations in the physical types in the Horseshoe Lake flock, and they are probably common to all goose populations. (4) Sexual dimorphism in weight and size in *maxima* exceeds that of any other race of *canadensis* (Table 5 and Hanson, *unpublished*); consequently, adult males of extreme weight diverge from the average for the other age-sex classes of *maxima* by a greater per cent than that which is found in the other subspecies.

5. Sexual dimorphism in body weight and wing length in populations of *Branta canadensis maxima, B. c. interior, B. c. moffitti, and B.c. canadensis*

Population	Difference in body weight (per cent larger, males)	Difference in wing length (per cent larger, males)
B. C. MAXIMA		
Round Lake, Minn.	18.3	7.9
Southern Manitoba	25.3	6.7
Rochester, Minn.	26.3	5.7
B. C. INTERIOR		
Horseshoe Lake, Ill.	14.3	6.4
B. C. MOFFITTI		
Museum skins	—	5.3
B. C. CANADENSIS		
Museum skins	—	5.6

BODY PROPORTIONS

The giant Canada goose is characterized by its long neck which in proportion to its body length exceeds that of all the other races. Thompson (1891), who had observed a brood raised in captivity near Shoal Lake, Manitoba, was prompted to refer to them as "swan-like birds." The long neck of this race is apparent in Figures 1, 9, 10, and 13. It is the only race which has a neck long enough to form a complete U when bent back on itself. In order that the reader can make his own comparisons in regard to this characteristic and others to be discussed, the photographs of the three other large races are included (Figs. 3–5).

WING SPAN AND BODY LENGTH

Next to weight, the wing span of the giant Canada goose most dramatically emphasizes its large size and is the linear dimension most commonly taken by the layman. Under the beguiling title "The Largest Goose on Record" in the *American Sportsman* (Anon., 1874), a goose taken on the Lime River near Benson Grove, Winnebago County, Iowa, April 19, 1874, was reported to have a wing span of 75 inches and a body length of 48 inches. The wing span of this bird is matched by those of living specimens, a number of which I measured for comparative purposes.

The largest geese I measured at the Round Lake Waterfowl Station had wing spans of 76½, 76½, 76, 74½, and 71½ inches. One of the geese with a 76½-inch span had a flattened wing measurement of 566 mm (22¼ inches), the longest of any measured in this study. At the North Dakota School for the Deaf, Devils Lake, a captive with a wing span of 76¾ inches was measured (flattened wing measurement of 555 mm). At Delta, Manitoba, the goose with the greatest length of flattened wing had a span of 73 inches. An 18¾-pound goose killed in Oakes, North Dakota, in 1947, was reported to have a wing span of 75¾ inches (Fig. 8). However, most of the large adult males in the various populations of *maxima* have wing spans between 69 and 71 inches. In contrast, large adult males of *interior* have a wing span of about 66 inches.

There are, however, records of two geese with wing spans exceeding any of the above measurements. A. J. Poeppe, who has a captive flock near Emmettsburg, Iowa, reported in personal conversation that he once had a Canada goose with a wing span of 84 inches. I have accepted this extreme record as authentic. However, a goose with a wing span of 88 inches and weighing 24 pounds has been reported in California (*see* Chapter 4, "Migration"). Is the report of a goose of this size from California just fiction? My first reaction was to consider it as such, but after learning of the 84-inch record from Iowa, I am less skeptical.

WING LENGTH

Differences in wing length between the various populations of *maxima* are much less than are the differences in body weight, a finding to be expected as weight is more closely correlated with the cube of linear measurements (Tables 4 and 6). The averages of length of wing for the various age-sex classes in the Rochester flock and for those in the southern Manitoba flocks are in excellent agreement with the averages of museum skins (Table 7). This is additionally significant if one considers that the latter includes both yearlings and adults. The Round Lake population of *maxima* is, however, distinctly longer-winged than any of the other populations of *maxima* measured. The largest museum skin measured, an adult male, had a wing length of 542 mm. It is noteworthy that this specimen was from a population breeding in Canyon County, Idaho, which available information indicates may be *maxima* (Tables 12 and 13). The Idaho population is discussed in Chapter 3, "Breeding Range."

The four large races constitute a graded series in respect to wing length — *maxima* > *moffitti* > *interior* > *canadensis*. Using two times

6. Length of wing in millimeters of *Branta canadensis maxima*, *B. c. interior*, *B. c. moffitti*, and *B. c. canadensis*

Subspecies by age-sex class	Number	Average	Range	Standard deviation
IMMATURE MALES				
B. c. maxima[1]	8	521.4 ± 3.0	507–537	8.6
B. c. maxima[2]	19	495.3 ± 3.5	480–520	15.3
B. c. maxima[3]	21	493.0 ± 3.2	461–519	14.8
B. c. interior[4]	114	485.7 ± 1.2	438–515	13.1
B. c. interior[5]	25	477.3 ± 1.9	463–495	9.5
YEARLING MALES				
B. c. maxima[3]	15	508.3 ± 4.2	481–535	16.3
B. c. interior[5]	15	499.5 ± 4.8	455–531	18.6
ADULT MALES				
B. c. maxima[1]	8	539.9 ± 6.0	510–566	17.0
B. c. maxima[2]	10	528.2 ± 4.0	510–547	12.6
B. c. maxima[3]	17	517.8 ± 3.9	485–542	16.0
B. c. interior[5]	43	509.9 ± 1.9	485–542	12.2
ADULT AND YEARLING MALES				
B. c. maxima[6]	16	511.1 ± 3.8	487–533	15.3
B. c. interior[4]	110	506.5 ± 1.3	463–547	13.6
B. c. moffitti[6]	13	516.6 ± 3.2	486–532	11.6
B. c. canadensis[6]	15	494.3 ± 3.3	476–518	12.8
IMMATURE FEMALES				
B. c. maxima[1]	3	500.0 ± 11.6	480–520	20.0
B. c. maxima[2]	8	473.8 ± 1.9	470–481	5.4
B. c. maxima[3]	12	475.8 ± 3.4	463–499	11.7
B. c. interior[4]	98	460.5 ± 1.5	410–488	14.7
B. c. interior[5]	25	457.1 ± 2.9	431–480	14.4
YEARLING FEMALES				
B. c. maxima[3]	14	478.6 ± 3.8	450–499	14.3
B. c. interior[5]	22	471.6 ± 3.0	455–499	14.0
ADULT FEMALES				
B. c. maxima[1]	14	500.5 ± 4.0	470–525	14.8
B. c. maxima[2]	7	495.0 ± 5.1	472–513	13.4
B. c. maxima[3]	11	489.9 ± 5.0	475–509	16.6
B. c. interior[5]	43	479.1 ± 1.6	457–509	10.8
ADULT AND YEARLING FEMALES				
B. c. maxima[6]	5	494.4 ± 5.0	485–512	11.0
B. c. interior[4]	92	481.4 ± 1.4	453–523	13.3
B. c. moffitti[6]	7	490.7 ± 3.8	476–505	10.3
B. c. canadensis[6]	12	468.2 ± 4.0	445–495	14.1

1. Captives measured at the Round Lake Waterfowl Station, Round Lake, Minnesota. 2. Includes samples from three southern Manitoba localities: The Delta Waterfowl Research Station, East Meadow Ranch, Oak Point, and the Alf Hole Provincial Refuge, Rennie. 3. Wintering geese at Rochester, Minnesota. 4. Wintering geese, Horseshoe Lake, Illinois (*from* Hanson, 1951). In retrospect, I realize that the series probably included a few individuals of the smaller, undescribed race which winters mainly near St. Marks, Florida. 5. Wintering geese, Horseshoe Lake, Illinois, 1957–62. 6. Museum skins.

7. Measurements in millimeters of museum skins of *Branta canadensis maxima* examined

Age-sex class	Locality
MALES	
Immature	Nebraska: Hall Co., Wood River
Immature	Nebraska: Platte River (county unknown)
Immature[1]	North Dakota: Towner Co.
Average	
Adult	Saskatchewan: Osler
Adult	Saskatchewan: Lost Mountain Lake
Adult	Saskatchewan: 10 mi. south of Kinistino
Adult	North Dakota: Kidder Co., Dawson
Adult	North Dakota: Ramsey Co.
Adult[2]	Minnesota: Lac Qui Parle Co., Madison
Adult[3]	Minnesota: Marshall Co.
Adult[4]	Minnesota: Grant Co., Round Lake
Adult	Nebraska: Hall Co., Wood River
Adult	Kansas: Lyon Co., Emporia
Adult[5]	Michigan: Schoolcraft Co., Seney Nat. Wildl. Refuge
Adult	Illinois: Cook Co., Chicago
Adult	Missouri: Macon Co.
Adult[6]	—
Adult[7]	? (Rothschild collection)
Adult[7]	? (Rothschild collection)
Average	
FEMALES	
Immature[8]	Saskatchewan: Lost Mountain Lake
Immature	Minnesota: Marshall Co., Mud Lake
Average	
Adult	Ontario: Lake St. Clair
Adult	Manitoba: Winnipeg
Adult	North Dakota: Ramsey Co., Sweetwater Lake
Adult	North Dakota: Nelson Co., Stump Lake
Adult	Minnesota: Marshall Co., Thief Lake
Average	

1. Shown in Figure 22. 2. Weight, 11 pounds. 3. Has distinct white spots over each eye. 4. Type specimen: has prominent white spot across forehead, white neck ring 18–20 mm wide, a yearling bird. 5. Pinioned bird; probably one

the standard error of the mean as a criterion, interracial differences are significant except between yearlings and adult females of *interior* and *moffitti* (Table 6).

TAIL LENGTH

The length of the tail is a useful measurement in separating *interior* from *maxima* in the immature age classes as the differences are significant. In the adults, however, only the largest-sized stocks of *maxima* can be separated from the other races by this measurement (Table 8). Again,

Date	Wing	Tail	Culmen Length	Culmen Width	Middle toe and claw	Tarsus	Museum
Oct. 31, 1884	480	121	61	23.7	101+	98	USNM
— 1900	—	132	54	23.2	99	—	USNM
— 1895	511	156	63	25.2	102	—	CNHM
	495.5	136.3	59.3	24.0	100.7	98.0	
Sept. 15, 1893	500	150	59	23.2	92	98	MCZ
April 29, 1932	519	148	59	—	93	99	USNM
May 29, 1962	491	153	57	—	95	97	USNM
Nov. 25, 1922	514	152	61	24.7	107	—	USNM
Nov., 1901	509	153	57	22.5	94	91	CNHM
April 11, 1892	496	158	61	23.5	—	—	MMNH
June 30, 1889	510	145	59	23.2	—	—	AMNH
April 22, 1876	528	140	68	26.8	—	—	AMNH
Nov. 4, 1884	533	155	61	25.2	101	—	USNM
March, 1891	530	161	61	23.8	92	98	MCZ
1936	521	155	65	25.6	99	101	USNM
Nov. 1864	535	151	60	23.7	104	—	PANS
March 20, 1875	498	155	68	25.5	101	97	USNM
—	487	147	58	23.0	—	—	AMNH
—	501	154	53	23.3	95	—	AMNH
—	505	154	53	22.9	—	—	AMNH
	511.1	151.9	60.0	24.0	98.3	97.3	
April 29, 1932	469	142	55	—	89	—	USNM
June 18, 1901	475	133	57	22.0	—	—	MMNH
	472.0	137.5	56.0	22.0	89	—	
April 25, 1884	495	150	55	—	87	—	ROM
April 13, 1906	485	148	57	23.3	90	93	CNHM
April 21, 1896	512	153	61	24.5	93	94	MCZ
Sept. 1, 1901	495	148	56	—	100	95	MCZ
May 2, 1891	485	139	55	22.6	92+	91	CNHM
	494.4	147.6	56.8	23.5	92.4	93.3	

of the original flock released. 6. Originally from the collection of J. J. Audubon. 7. Had been mounted and used for exhibition. 8. Has adult tail feathers.

the four large races constitute the same graded series in size — *maxima* > *moffitti* > *interior* > *canadensis* — although most of the interracial differences for this measurement are not statistically significant.

THE BILL

The bill of *maxima* differs from the other large races by its spatulate, relatively untapered shape (Figs. 14–16). In addition, most examples of *maxima* can be distinguished by feeling the horny palate of the upper bill. The palate of *maxima* is relatively smooth, with few protuberances;

8. Length of tail in millimeters of *Branta canadensis maxima, B. c. interior, B. c. moffitti,* and *B. c. canadensis*

Subspecies by age-sex class	Number	Average	Range	Standard deviation
IMMATURE MALES				
B. c. maxima[1]	9	153.9 ± 2.7	142–165	8.2
B. c. maxima[2]	12	140.1 ± 2.0	131–152	7.1
B. c. interior[4]	111	128.8 ± 0.6	115–149	6.2
YEARLING MALES				
B. c. interior	20	148.4 ± 1.4	142–163	6.2
ADULT MALES				
B. c. maxima[1]	8	170.6 ± 2.9	165–183	8.2
B. c. maxima[2]	2	171.0 ± 6.5	164–177	9.2
B. c. maxima[3]	14	156.5 ± 2.1	139–172	8.0
B. c. interior[4]	20	151.8 ± 1.5	138–160	6.8
ADULT AND YEARLING MALES				
B. c. maxima[5]	16	151.9 ± 1.3	140–161	5.1
B. c. interior[4]	109	150.6 ± 0.6	135–165	6.1
B. c. moffitti[5]	13	152.5 ± 3.4	124–165	12.1
B. c. canadensis[5]	15	144.1 ± 2.3	122–153	8.9
IMMATURE FEMALES				
B. c. maxima[1]	3	149.0 ± 6.6	137–160	11.5
B. c. maxima[2]	8	134.3 ± 1.8	125–141	5.2
B. c. interior[4]	98	122.2 ± 0.7	105–139	7.2
YEARLING FEMALES				
B. c. interior[4]	15	141.1 ± 1.3	129–150	5.1
ADULT FEMALES				
B. c. maxima[1]	14	157.7 ± 1.8	145–166	6.9
B. c. maxima[2]	1	145	—	—
B. c. maxima[3]	7	148.0 ± 3.0	136–159	8.0
B. c. interior[4]	20	146.1 ± 1.3	135–153	5.6
ADULT AND YEARLING FEMALES				
B. c. maxima[5]	5	147.6 ± 2.4	139–153	5.2
B. c. interior[4]	90	142.4 ± 0.7	132–162	6.5
B. c. moffitti[5]	6	143.2 ± 3.4	134–157	8.6
B. c. canadensis[5]	12	138.5 ± 1.9	125–150	6.6

1. Captives measured at the Round Lake Waterfowl Station, Round Lake, Minnesota. 2. Includes samples from three southern Manitoba localities: Delta Waterfowl Research Station, East Meadow Ranch at Oak Point, and the Alf Hole Provincial Refuge, Rennie. 3. Wintering geese at Rochester, Minnesota. 4. Wintering geese at Horseshoe Lake, Illinois (series between 90 and 111 *from* Hanson, 1950). 5. Museum skins.

in the other races, the horny palate is narrower, more troughlike, and the protuberances are prominently arranged in rows and fairly sharp to the touch. The horny palate of *maxima* is also usually much less pigmented than that of most individuals of the other subspecies.

The nail of the bill is another distinguishing characteristic; in *maxima* it is rounder and more bulbous at the tip than in the other races and it tends to cup around the lower mandible to a greater extent. Viewed in profile, the bill is less tapered than in the other races, and in most individuals it appears fairly blunt, reminiscent of decoys. In contrast, the nails of the bills of *B. c. interior, canadensis,* and *moffitti* tend to be more triangular in shape, better defined, and merge less smoothly with the adjacent superior portion of the bill.

In *maxima,* the lamellae are especially coarse and their lateral termini are usually prominently exposed as "teeth" along the entire edge of the upper mandible (Fig. 17). The possible significance of the prominently serrated bill of *maxima* is discussed in Chapter 9, "Foods and Feeding Habits."

Data on the length of the exposed culmen of populations of *maxima* and the other three large races are summarized in Table 9. The averages for the samples of *maxima* males from Rochester and southern Manitoba do not differ significantly from the average of skins of males in museums (Table 7) or from what I consider the "gold standard" of *maxima* — the population at the Waubay National Wildlife Refuge, South Dakota. All are significantly different from the Horseshoe Lake specimens. Again, the geese from Round Lake, Minnesota, and the captives at the Des Plaines Wildlife Area in Illinois appear to fall into a distinct category as the averages for the various age-sex samples of these geese differ significantly from the other populations of *maxima*. It is noteworthy that the type of *maxima,* the largest specimen in the series of museum skins examined by Delacour or me, has a culmen of 68 mm. In January, 1963, the culmen measurement of this specimen was equaled by an adult male I measured at Rochester, Minnesota.

The massiveness of the bill of *maxima* is one of the most consistent and salient characters of the race. A comparison of this character of *maxima* with the other races is provided by measurements of bill width (Table 10). The relative width of the bill of the large races can also be seen in Figures 14–16. The bill of adult females of all of the large races including *maxima* is, however, more tapered laterally and the nail more pointed than in adult males. This factor must be taken into consideration in using bill shape and width as diagnostic characters.

9. Length of exposed culmen in millimeters of *Branta canadensis maxima, B. c. interior, B. c. moffitti,* and *B. c. canadensis*

Subspecies by age-sex class	Number	Average	Range	Standard deviation
IMMATURE MALES				
B. c. maxima[1]	9	64.7 ± 1.2	61–70	3.7
B. c. maxima[2]	14	64.2 ± 1.3	54–70	4.8
B. c. maxima[3]	24	58.1 ± 0.6	53–61	3.0
B. c. maxima[4]	18	58.1 ± 0.7	53–66	3.1
B. c. interior[5]	114	53.4 ± 0.2	47–61	2.5
B. c. interior[6]	13	54.5 ± 0.6	49–58	2.3
YEARLING MALES				
B. c. maxima[4]	8	57.5 ± 0.7	54–59	2.1
B. c. interior[6]	5	54.2 ± 0.04	53–55	0.1
ADULT MALES				
B. c. maxima[1]	8	65.3 ± 1.4	61–72	4.0
B. c. maxima[3]	10	58.3 ± 1.1	51–62	3.5
B. c. maxima[4]	9	59.4 ± 1.0	54–62	3.1
B. c. interior[6]	18	55.5 ± 0.6	51–60	2.6
ADULT AND YEARLING MALES				
B. c. maxima[7]	13	58.6 ± 0.8	55–63	2.9
B. c. maxima[8]	16	60.0 ± 0.7	53–68	1.4
B. c. interior[5]	110	53.7 ± 0.3	49–61	2.8
B. c. moffitti[8]	14	54.6 ± 0.6	50–58	2.1
B. c. canadensis[8]	15	56.1 ± 0.8	52–62	3.1
IMMATURE FEMALES				
B. c. maxima[1]	3	59.0 ± 1.8	57–63	3.1
B. c. maxima[2]	22	59.0 ± 0.7	52–64	3.3
B. c. maxima[3]	12	53.7 ± 0.8	49–57	2.8
B. c. maxima[4]	11	53.8 ± 0.9	49–57	3.0
B. c. interior[5]	98	50.2 ± 0.3	43–53	2.5
B. c. interior[6]	11	50.1 ± 0.7	48–53	2.3
YEARLING FEMALES				
B. c. maxima[4]	6	52.7 ± 0.7	51–55	1.6
B. c. interior[6]	11	51.3 ± 0.6	46–53	1.9
ADULT FEMALES				
B. c. maxima[1]	14	59.8 ± 0.9	55–63	3.4
B. c. maxima[3]	6	54.8 ± 0.5	53–56	1.2
B. c. maxima[4]	6	54.0 ± 2.0	51–56	5.0
B. c. interior[6]	19	49.9 ± 0.5	45–55	2.2
ADULT AND YEARLING FEMALES				
B. c. maxima[7]	8	53.5 ± 0.6	51–56	1.7
B. c. maxima[8]	5	56.8 ± 1.1	55–61	2.5
B. c. interior[5]	92	49.8 ± 0.3	43–56	2.4
B. c. moffitti[8]	10	51.6 ± 0.8	47–56	2.5
B. c. canadensis[8]	11	51.0 ± 0.8	45–55	2.8

1. Captives measured at the Round Lake Waterfowl Station, Round Lake, Minnesota. 2. Captive stock from Des Plaines Wildlife Area, Illinois. 3. Includes samples from three southern Manitoba localities: Delta Waterfowl Research Station, East Meadow Ranch, and the Alf Hole Provincial Refuge, Rennie. 4. Wintering geese at Rochester, Minnesota. 5. Wintering geese, Horseshoe Lake, Illinois (*from* Hanson, 1950). 6. Wintering geese, Horseshoe Lake, Illinois, 1957–62. 7. Flock at Waubay National Wildlife Refuge, Waubay, South Dakota. (Data courtesy of Robert R. Johnson and Ray D. Hart.) 8. Museum skins.

10. Width of culmen in millimeters at midpoint of nares of *Branta canadensis maxima, B. c. interior, B. c. moffitti,* and *B. c. canadensis,* yearlings and adults combined

Sex and subspecies	Number	Average	Range	Stand-ard devi-ation	Source of specimens measured
MALES					
B. c. maxima	19	24.6 ± 0.2	23.0–25.7	0.8	Rochester, Minnesota
B. c. maxima	14	24.0 ± 0.6	22.9–26.8	2.3	Museum skins[1]
B. c. interior	22	22.8 ± 0.2	21.0–24.1	0.8	Horseshoe Lake, Ill.
B. c. moffitti	10	22.2 ± 0.1	21.7–22.9	0.4	Museum skins
B. c. canadensis	14	22.1 ± 0.2	21.1–23.3	0.7	Museum skins
FEMALES					
B. c. maxima	10	23.8 ± 0.3	22.3–25.4	1.0	Rochester, Minnesota
B. c. maxima	3	23.5 ± 0.2	22.6–24.5	0.3	Museum skins
B. c. interior	20	21.4 ± 0.1	20.7–22.7	0.6	Horseshoe Lake, Ill.
B. c. moffitti	6	21.0 ± 0.3	20.0–21.8	0.7	Museum skins
B. c. canadensis	11	21.1 ± 0.3	20.2–22.4	0.3	Museum skins

1. Culmens of museum skins would have undergone a slight amount of shrinkage.

THE SKULL

The basic qualitative differences between the skull of *maxima* and the other races, as exemplified by *interior*, are well shown in Figure 18. The most prominent difference, other than size, is the untapered shape of the bill of *maxima*, particularly the nail portion.

THE TARSUS AND FOOT

SCUTELLATION — The scutellation of the tarsi and feet of *maxima*, especially the tarsi, is distinctive; for this reason alone, it is surprising that the race went undescribed for so long. In essence, the skin of the tarsi of *interior, moffitti,* and *canadensis* is suggestive of a smooth-skinned colubrid snake; in *maxima*, the scutes are more plaque-like, their central portions are depressed, and the grooves between the scutes are deeper and more pronounced. The scutes of *maxima* are also larger and rounder than in the other races, and the pattern of their arrangement is less regular (Fig. 19). The net effect is that the leg of *maxima* has a "pachy-dermal" appearance (Figs. 1 and 19).

The color of the skin varies with age and with the season of the year, being olive-toned in goslings and also in adults during the molt. In winter, the skin of the leg of the large descendants of decoy stocks of *maxima* tends to be grayish; that of the giant Canadas at Rochester in winter is black. The skin of the legs and feet of *interior* has a dark olive-brown or brownish-black cast.

LENGTH OF MIDDLE TOE — Another of the distinguishing characteristics of *maxima* is the large size of its feet relative to wing length. The foot of *maxima*, as shown by measurements of the middle toe, is the longest of the races of *canadensis* (Hanson, *unpublished*). Unlike the tarsus, the middle toe can be readily measured with satisfactory accuracy in living or freshly killed specimens. Measurements in Table 11 are from geese in both of these states. Only an approximate measurement of toe length can be made from museum skins (Table 7) as the feet are usually dried in a partly contracted condition.

Previous comments regarding wing length and length of exposed culmen for the various populations studied hold equally true for length of middle toe. A sample of *maxima* in Illinois, the origin of which is discussed in Chapter 3, "Breeding Range," averaged somewhat larger than the Round Lake geese in toe length, and probably would do so for all skeletal measurements if they were available. Both samples well exceed the average foot size of museum specimens of *maxima*; the latter are in the same size range as the Rochester and Manitoba specimens.

COLORATION OF THE ADULTS

In his original description of *maxima*, Delacour (1951:5) wrote:

> *Differs from B. c. canadensis* in its larger size and more elongate shape, and in its paler, more even plumage, less conspicuously barred above; the under parts more uniform, the base of the hind neck and upper back not whiter than the rest; a white ring at the base of the neck often present (as in type specimen).

In his monograph, "The Waterfowl of the World," (1954:164) Delacour gives the following comparative description of the race *moffitti:*

> Completely similar in colour to *B. c. maxima*, but less elongate, rounder in shape, with shorter bill, legs, and feet. Slightly bigger than *canadensis.*

Aldrich (1946:196) originally described the race *moffitti:*

> Similar to *Branta canadensis canadensis*, but larger and paler in general coloration. In fresh autumn specimens the brown areas of upper parts and flanks are between olive brown and buffy brown, rather than mummy brown. Underparts average somewhat less whitish.

11. Length of middle toe and claw in millimeters of *Branta canadensis maxima, B. c. interior, B. c. moffitti,* and *B. c. canadensis*

Subspecies by age-sex class	Number	Average	Range	Standard deviation
IMMATURE MALES				
B. c. maxima[1]	9	106.2 ± 1.0	101–111	2.9
B. c. maxima[2]	14	110.8 ± 2.5	104–116	9.4
B. c. maxima[3]	21	97.5 ± 0.8	90–101	3.5
B. c. interior[4]	15	92.8 ± 1.2	85–100	4.5
ADULT MALES				
B. c. maxima[1]	8	106.5 ± 0.9	104–112	2.6
B. c. maxima[3]	10	100.3 ± 1.7	88–108	5.5
B. c. maxima[5]	15	101.6 ± 1.1	92–108	4.4
B. c. interior[4]	25	95.0 ± 0.6	91–100	2.8
YEARLING AND ADULT MALES				
B. c. maxima[6]	12	98.3 ± 1.4	92–107	5.0
B. c. moffitti[6]	8	94.6 ± 1.4	86–100	3.9
B. c. canadensis[6]	7	90.1 ± 1.1	86–94	3.1
IMMATURE FEMALES				
B. c. maxima[1]	3	103.0 ± 2.0	99–105	3.5
B. c. maxima[2]	22	100.3 ± 0.9	94–110	4.0
B. c. maxima[3]	8	92.0 ± 1.3	86–98	3.7
B. c. interior[5]	11	87.9 ± 1.0	81–93	3.3
ADULT FEMALES				
B. c. maxima[1]	13	98.9 ± 1.6	90–110	5.9
B. c. maxima[3]	6	90.0 ± 1.1	87–95	2.8
B. c. maxima[5]	11	94.1 ± 1.4	88–105	4.8
B. c. interior[4]	20	89.0 ± 0.5	84–92	2.3
ADULT AND YEARLING FEMALES				
B. c. maxima[6]	5	92.4 ± 3.0	87–100	6.8
B. c. canadensis[6]	3	87.3 ± 0.1	87–88	0.2

1. Captives measured at the Round Lake Waterfowl Station, Round Lake, Minnesota. 2. Captive stock from Des Plaines Wildlife Area, Illinois. 3. Includes samples from three southern Manitoba localities: Delta Waterfowl Research Station, East Meadow Ranch at Oak Point, and the Alf Hole Provincial Refuge, Rennie. 4. Wintering geese at Horseshoe Lake, Illinois. 5. Wintering geese at Rochester, Minnesota. 6. Museum skins.

This description, however, is marred by its being based in part on specimens of *maxima.* It may have been this factor that prompted Conover (Hellmayr and Conover, 1948:303) to remark:

> The series in Field Museum [Chicago Natural History Museum] shows that as one goes west from the Mississippi Valley these geese [*moffitti*] gradually get lighter on the breast, but that there is great individual variation and that

even so the California birds are darker than typical *canadensis* from the east coast of North America.

Coloration in *maxima,* as the race is construed here, is however a more complex problem than the above authors have indicated.[2] Specimens of *maxima* collected in January at Rochester, Minnesota, are, superficially, nearly indistinguishable from a series of specimens of *moffitti* collected in the fall in Utah if sex differences are not taken into consideration. The differences in coloration between the sexes are, however, somewhat greater in *maxima* than in *moffitti.* The males of *maxima* (yearlings and adults) are lighter in color on the breast than *moffitti,* the former being a somewhat more pearly gray. Yearling and adult female *maxima,* on the other hand, may have breast color nearly as dark as *interior* (Fig. 20).

But the most notable color differences between the Rochester *maxima* and the Utah *moffitti* are shown by the wings. The upper wing coverts of the Rochester *maxima* have a somewhat more grayish cast than *moffitti,* this difference being more decidedly expressed by the under wing coverts which are grayish-tan in *maxima* and a tannish-brown in *moffitti.* The outer portions of the secondary feathers of the wing are grayish-brown in the Rochester population of *maxima* and brown in *moffitti.* Even these differences do not provide absolute separation of the two races, although it was concluded from the specimens studied that it would be possible to separate 80 to 90 per cent of the individuals in a population of one race on the basis of color from a similar percentage in the other. (Size, body proportions, shape of bill, and scutellation and markings do, of course, readily permit separation of the Rochester and Utah populations.)

Paradoxically, large specimens of *maxima* from Denver, Colorado,[3] and from Round Lake, Minnesota, are more like *moffitti* in respect to the color of the breast and wings than are examples of the somewhat smaller stocks of *maxima* originating in Manitoba; they differ from the Manitoba populations by being darker, the back having a more mahogany or chestnut-brown color, closely approaching the back color of *interior* collected at Horseshoe Lake, Illinois. The race *maxima* also differs from *interior* by having a more extensively developed area of grayish-tan at the base of the neck below the black stocking. One of the Denver specimens is notable in this regard, having a very prominent light gray mantle over the forepart of the back and base of the neck. However, on February 22, 1964, I was able to match this extreme example with a live adult male in the Rochester flock (the skin of the Denver bird was at hand at

12. Measurements of museum skins in millimeters of Canada geese from Idaho judged to be most closely related to *Branta canadensis maxima*

Age-sex class	Locality	Date	Wing	Tail	Culmen Length	Culmen Width	Middle toe and claw	Museum
Adult male	Idaho: Canyon Co., Deer Flat Reservoir	Oct. 10, 1947	533	168	56	24.2	101	CM
Adult male	Idaho: Canyon Co., Deer Flat Reservoir	Oct. 10, 1947	542	163	56	22.9	101	CM
Average			537.5	165.5	56.0	23.6	101	
Adult female	Idaho: Canyon Co., Deer Flat Reservoir	Oct. 10, 1947	470	148	54	21.3	85+	CM
Adult female	Idaho: Canyon Co., Deer Flat Reservoir	Oct. 10, 1947	496	138	52	22.1	90+	CM
Adult female	Idaho: Canyon Co., Deer Flat Reservoir	Oct. 10, 1947	490	141	53	—	—	CM
Adult female	Idaho: Lewiston	Oct. 28, 1934	500	137	56	—	101	USNM
Average			489.0	141.0	53.8	21.7	92.0	

the time for comparison). Also, the upper wing coverts of the large *maxima* from Denver and from Round Lake, Minnesota, are horn-brown or gray-brown in comparison with *interior* in which these feathers are more reddish or chestnut-brown.[4] The significance of these variations is evaluated in Chapter 15, "Discussion."

THE CHEEK PATCHES — The white area of the cheek tends to be more extensive in *maxima* than in the other large races (Figs. 2–5). It often extends slightly farther down the neck and on the head, nearly to the crown level, giving it a wrap-around appearance. Although the distinctive shape of the bill is probably the best diagnostic character of the head, the presence of a small, often hooklike extension near the top of the posterior margin of the cheek patches may be regarded as an excellent indicator of a *maxima* population. It was commonly seen in the flocks at Rock County, Wisconsin; Rochester, Minnesota; Rennie, Manitoba; in the Strutz flock at Jamestown, North Dakota; in free-flying geese at Calgary, Alberta, and Fort Whyte, Manitoba; in captive stock at Bay Beach Park, Green Bay, Wisconsin; and in free-flying geese at the Waubay National Wildlife Refuge. It is exhibited by a specimen collected in 1895 in North Dakota (Figs. 10 and 22).

WHITE MARKINGS — Two white markings frequently occur in *maxima:* (1) a white spot or bar across the forehead (it may be limited to a spot above each eye), and (2) a pure white neck ring. However, in all the large races of *Branta canadensis* the occasional individual is seen with extensive white spotting on the head and neck (Hanson, 1949*b*) (Figs. 23–25).

THE FOREHEAD SPOT — The forehead spot is frequently found in *maxima* and can be regarded as a fairly definitive characteristic of the race (Fig. 10); it is quite rare in the other races of *canadensis*. Using a 20X spotting scope, Harvey K. Nelson, Forrest B. Lee, and Don Smith observed some degree of forehead spotting in 20 per cent of 1,121 geese examined in the Rochester flock in the winter of 1959–60. The type specimen in the American Museum of Natural History and two specimens in the Chicago Natural History Museum exhibit this marking (Fig. 22). It is a not uncommon marking of birds in the Rock County and Rochester flocks. It was seen in the flocks at Rennie (Fig. 1), Fort Whyte (Fig. 28), East Meadow Ranch (Fig. 29), and Delta, Manitoba, and, judging from photographs, is present in populations of the race nesting in Montana and Alberta (Figs. 30 and 31). It is a conspicuous characteristic of the original wild stocks in North Dakota (Figs. 22 and 35) and is exhibited by some of the birds in the flock owned by Carl E. Strutz of Jamestown, North Dakota (Fig. 9), and offsprings of this flock (Fig. 10). A female collected by Tom Sterling on the Thelon River in July, 1963, had a white forehead band (Fig. 25). (*See* Chapter 4, "Migration," for discussion of the significance of similar records.) The only flock of *maxima* in which it was not noted was the group of 85 immatures produced by the captive flock held by the Illinois Department of Conservation, although it occurred in some of the parent stocks of these birds.

13. Length of culmen in millimeters of Canada geese from two localities in Idaho

	B. c. maxima—Canyon County			B. c. moffitti—Caribou County		
Age-Sex class	Number	Average	Range	Number	Average	Range
Immature male	3	56.0	55–58	2	52.0	—
Adult male	6	56.0	53–63	2	52.5	51–54
Total or average	9	56.0		4	52.3	
Immature female	9	53.2	48–57	4	51.8	49–55
Adult female	2	55.0	—	6	53.2	51–58
Total or average	11	53.5		10	52.6	

The white spot on the forehead may be flecked with varying amounts of black; the tips of these feathers may be black but their bases are white. As Figgins (1920:99–100) has pointed out, the feathers of the head are subject to reduction in length as a result of wear. (This wear, for example, causes the feathers of the white cheek patches of *interior* to lose

1. *Giant Canada geese at the Alf Hole Provincial Refuge, Rennie, Man. Note swan-like neck of bird above and the white forehead spot of the goose below.*

2. (At left) *End of a successful hunt on the Great Plains,* 1931. *Geese shown are unquestionably giant Canada geese. Photograph by Otto M. Jones; reproduced from Hornaday* 1931, *courtesy of Charles Scribner's Sons.*

3. *Atlantic Canada geese,* Branta canadensis canadensis, *at Long Island* (*at the right*) *and Westchester County* (*below*), *N.Y. Photographs courtesy of Allan D. Cruickshank.*

4. (Above) *Hudson Bay Canada geese,* Branta canadensis interior, *at Horseshoe Lake, Alexander County, Ill. Adult male.* (Below) *Adult male* B. c. interior *on right; on left, adult male of a new race which breeds along the south coast of Baffin Island and on Southampton Island and which will be named by the author in honor of Jean Delacour.*

5. (Above and below) *Western Canada geese,* Branta canadensis moffitti, *at the Idaho State Game Farm, near Lewiston. These geese are typical of those breeding along the lower Snake River in Washington. Photographs courtesy of Charles F. Yocum.*

6. (Above, right) *The giant Canada goose shown on the right in this photograph weighed a few ounces over 21 pounds. It was shot in 1915 in Platte County, Mo.* Photograph courtesy of Robert A. Brown. (Below, left) *A 19-pound giant Canada goose shot at Horseshoe Lake, Alexander County, Ill., November 26, 1941.* Photograph courtesy of William Brown. (Below, right) *Four giant Canada geese shot in 1924 near Bismarck, N.D.* Photograph courtesy of Morris D. Johnson.

8. (Above) *An 18¾-pound Canada goose shot near Oakes, N.D., in late November, 1947.* Photograph courtesy of Kenneth Brossmann.

7. (Below, right) *Giant Canada goose on left weighed 19½ pounds and was shot in Walworth, Wis., December 25, 1924.* Photograph courtesy of Wisconsin Conservation Department.

9. (Above) An 18-pound male giant Canada goose at the Carl E. Strutz Game Farm, Jamestown, N.D. Note white, oval, forehead spot. (Below) A large male at the Lac Qui Parle State Game Refuge, Minn. Note large feet and rough scaling of tarsi. This goose originally came from the flock of E. E. Kern, Granada, Minn.

10. A giant Canada goose at Lower Souris National Wildlife Refuge. This is an immature male weighing over 14 pounds, a progeny of the large male at the Strutz Game Farm (Fig. 9). Note hooklike extension of posterior edge of cheek patch. Below note unusually extreme, upright stance. The ability to maintain this nearly upright posture is probably characteristic of maxima and a reflection of its distinctive body proportions. This posture observed in the race interior when they were momentarily stretching to reach corn on the stalk.

11. A part of a flock of giant Canada geese at the Oscar Luedtke farm, Lotts Creek, Ia.

12. A large male giant Canada goose at the Round Lake Waterfowl Station, Round Lake, Minn.

13. *Giant Canada geese at Silver Lake Park, Rochester, Minn. The two geese below are probably a mated pair; the darker bird on the left is undoubtedly the female. In the race interior, the adult males have a neck that, relative to body size, is 15 per cent longer than adult females (H. C. Hanson and H. I. Fisher, unpublished).*

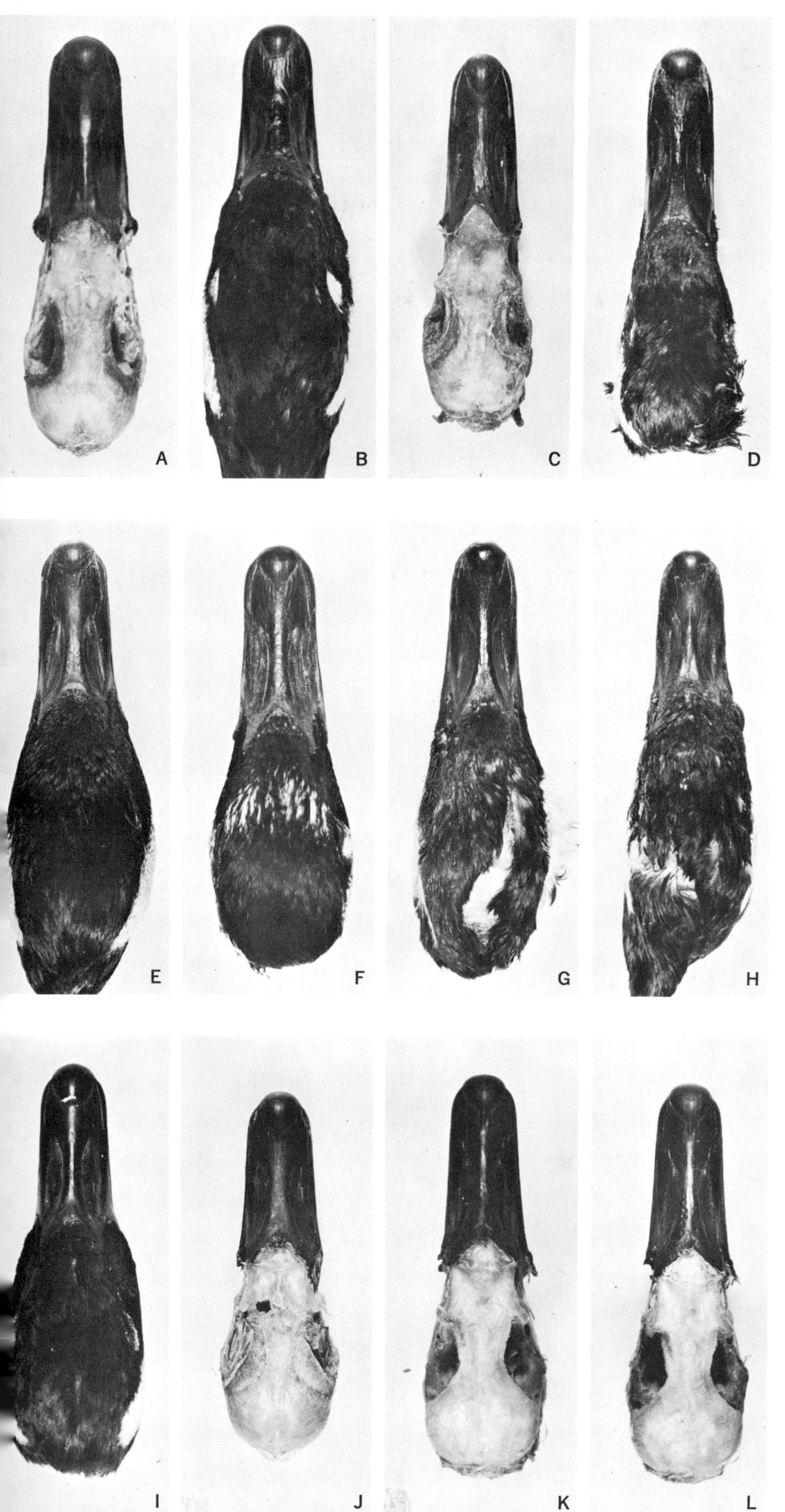

14. (Overleaf)

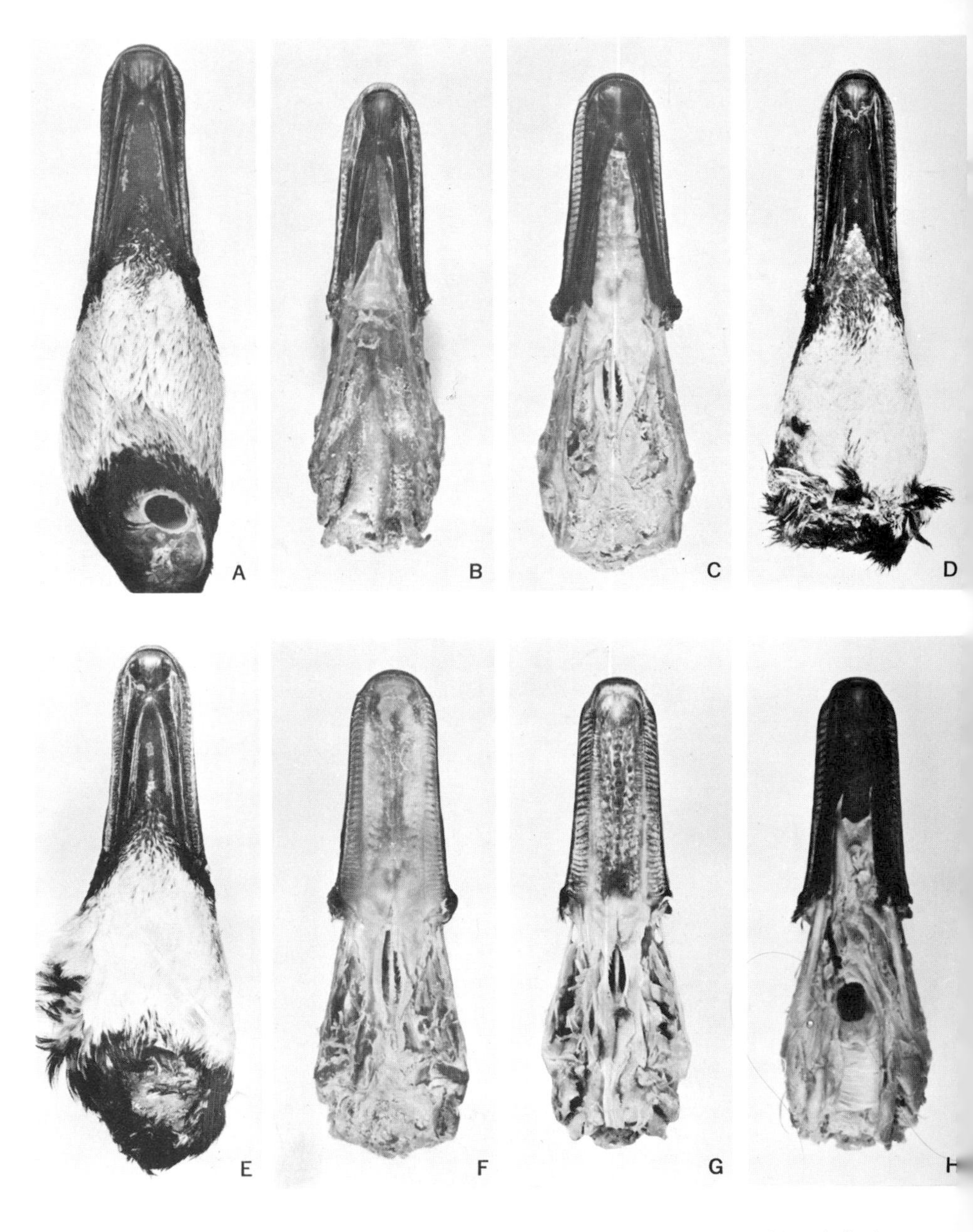

15. (Above) Ventral view of heads of male B. c. maxima: (A) adult from Swan Lake, Mo.; (B) from Dog Lake, Man.; (C) immature from Rochester, Minn.; (D) from Canyon County, Id.; and (E) yearling from Rochester, Minn. (G and H) Head of adult male B. c. interior from Horseshoe Lake, Ill. Note the differences between the horny palate of maxima (F) and interior (G).

14. (Overleaf) Heads of B. c. maxima: (A) immature male and (B) yearling male from Rochester, Minn.; (C) adult male from Dog Lake, Man. and (D) male (?) from the interlake area of Man.; (E) immature male and (F) adult male, both from Swan Lake, Mo.; and (G) adult male from Canyon County, Id. Heads of B. c. moffitti: (H) adult male from Blackfoot Reservoir, Id., and (I) adult male from Bear River marshes, Ut. (J) Head of B. c. canadensis from Chesapeake Bay, Md. (K and L) Heads of B. c. interior from Horseshoe Lake, Ill. Note that the anterior edge of the nail of maxima merges with the lateral portions of the culmen to make a smooth semicircle; in the other races (moffitti, canadensis, and interior) the nail projects anteriorly of the lateral portions of the culmen as a fairly distinct entity.

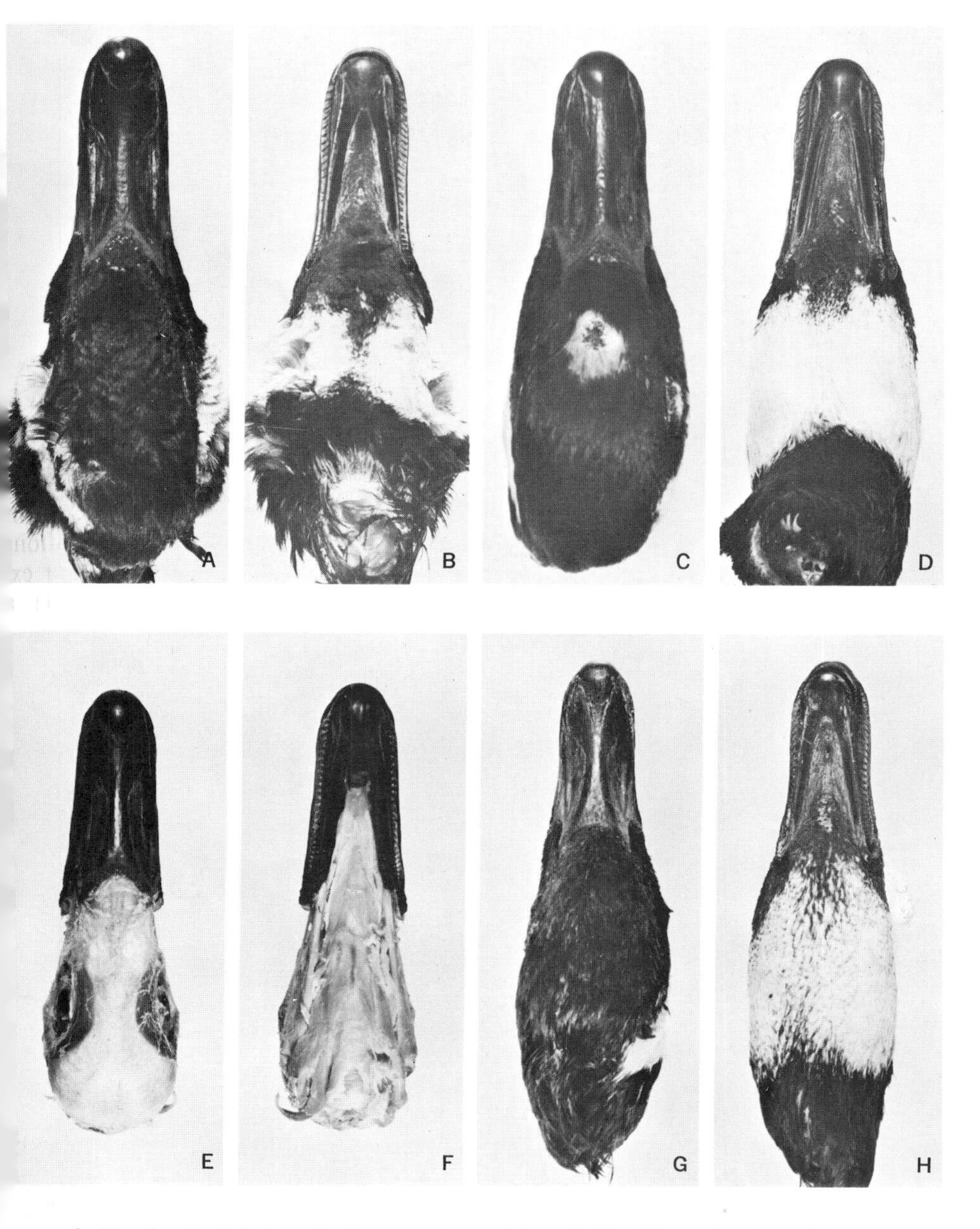

16. *Heads of adult female* B. c. maxima: *(A and B) old captive from Fayette County, Mo.; (C and D) adult female from Swan Lake, Mo.; and (E and F) from Rochester, Minn. (G and H) Head of an adult female* B. c. moffitti *from Blackfoot Reservoir, Id.*

17. (Left) *Heads of giant Canada gees shot at the Swan Lake National Wild life Refuge, autumn of 1962:* (A) *a 11¾-pound male; note serration along upper bill and the posterior ex tension of the cheek patch.* (B) *12½-pound female.*

18. *Skulls of* B. c. maxima: (A) *adult male from captive flock, Round Lake, Minn., and* (B) *adult male found dead at Horseshoe Lake, Ill., 1952;* (C *and* D) *adult males* B. c. interior *from Horseshoe Lake.*

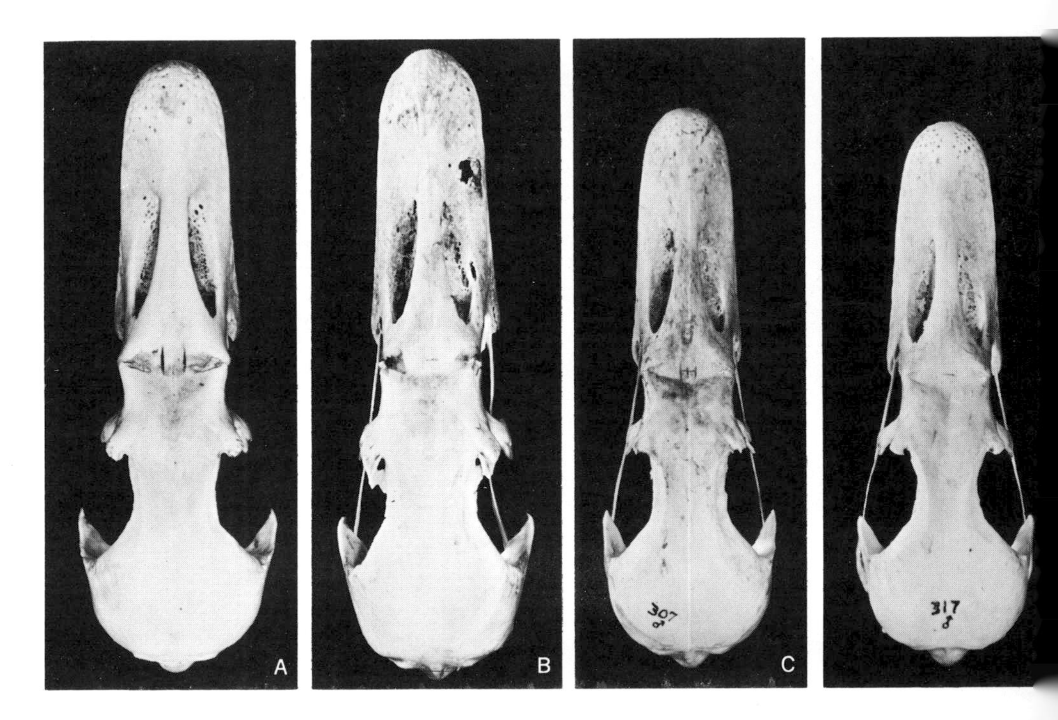

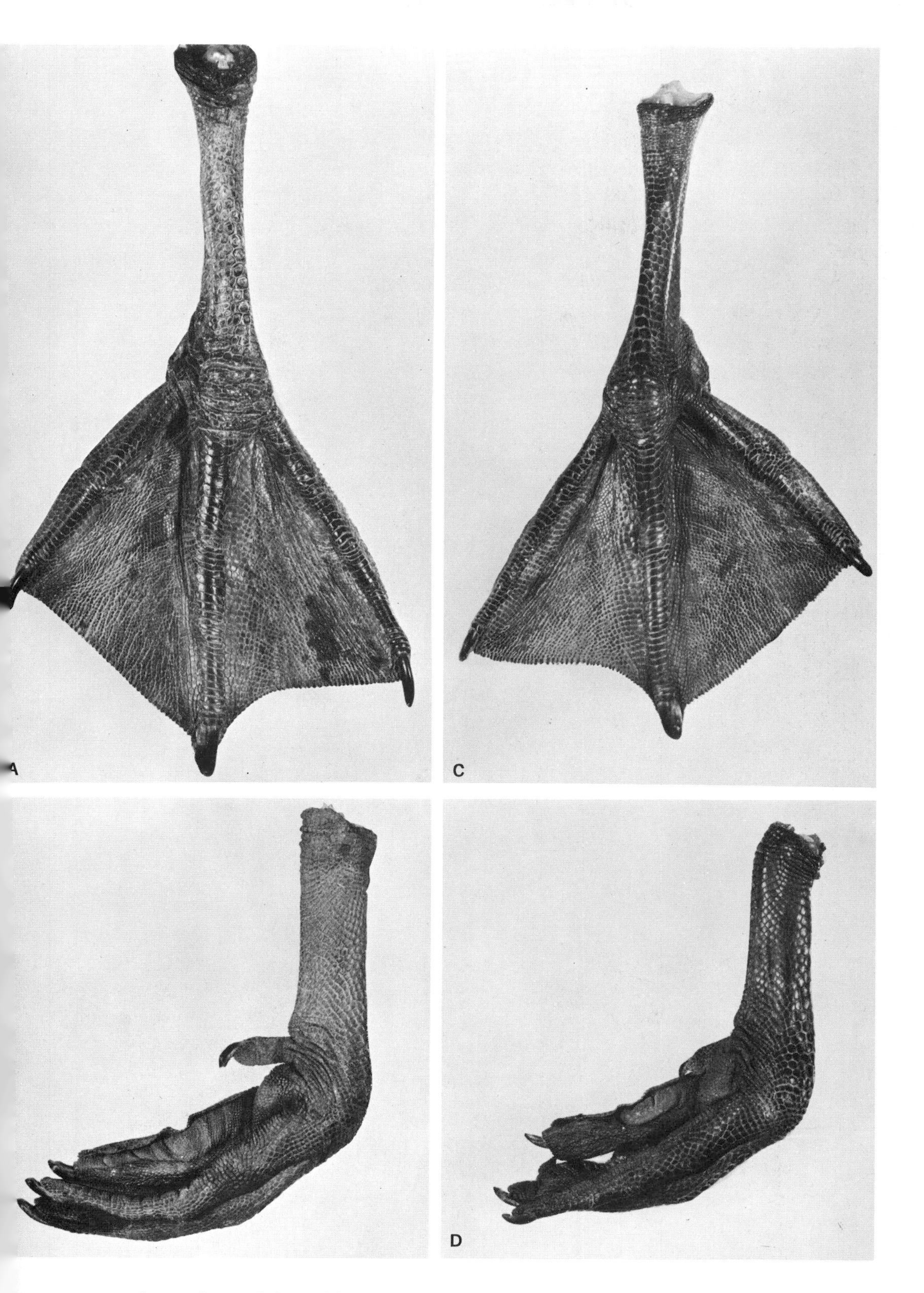

19. Lower leg and foot: (A and C) of B. c. maxima from an Illinois flock; (B and D) of B. c. interior, Horseshoe Lake, Ill. Note differences in size, shape, and arrangement of the scutes.

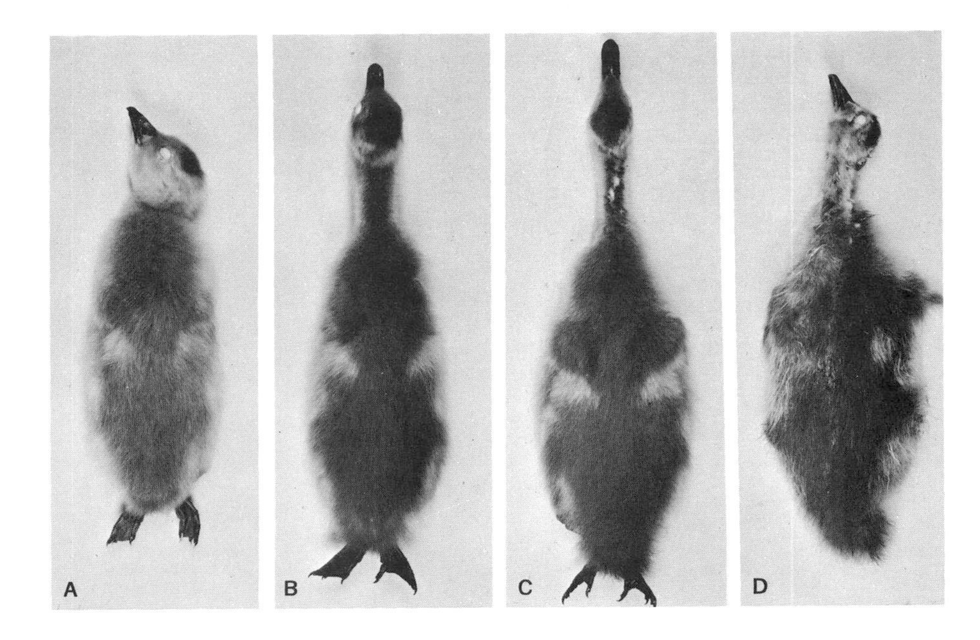

21. (Above) *Skins of gosling* Branta canadensis maxima: (*A*) *from Towner County, N.D., June 7, 1895;* (*B and C*) *from Seney Refuge, Mich., June 3, 1938, and June 5, 1949, respectively.* (*D*) *Skin of gosling* B. c. interior *from Akimiski Island, July 9, 1958.*

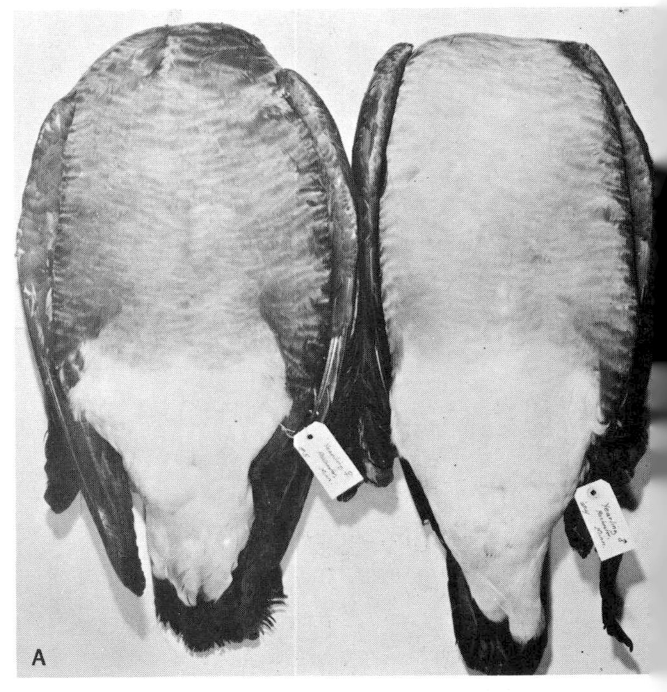

20. (Right, A) *Skins of 1½-year-old giant Canada geese collected at Rochester, Minn.; female and male.* (*Below, B*) *Adult female.*

their black speckling characteristic of the winter period and to become white by the onset of the breeding season.) Thus, some individuals of *maxima* develop an extensive white oval spot as a result of feather wear (Fig. 22). An extreme example in which the development of the white forehead patch is so extensive that it is confluent with the white cheek patches is shown in Figure 25. The pattern of pigmentation of the head feathers of this goose is reminiscent of the barnacle goose (*Branta leucopsis*).

THE NECK RING — Geese with a neck ring of pure white feathers at the base of the black stocking are common in most populations of *maxima*; in the wintering flock at Rochester, Minnesota, a white ring of varying width was present in over 50 per cent of the immatures, 62 per cent of the yearlings, and 82 per cent of the adults (Table 14 and Fig. 20). The neck ring may vary in width from as much as ½ to 1½ inches.

Comparable data collected for *interior* at Horseshoe Lake in February, 1962, just subsequent to the observations made at Rochester, revealed that neck rings were much less prevalent in this race (Table 14). In both races, particularly in *maxima*, the neck ring is more extensively developed in adults. However, with few exceptions, one aspect of the neck ring probably permits the two races to be separated: in *maxima*, the feathers of the neck ring are pure white and contrast sharply with the contour feathers of the underparts; in *interior*, the feathers of the neck ring are usually dusky white.

14. Prevalence of white neck rings in *Branta canadensis maxima* at Rochester, Minnesota, and *B. c. interior* at Horseshoe Lake, Illinois

		Relative size of neck ring				
		0	1	2	3	4
Subspecies by age class	Number examined	Frequency (per cent)	Frequency (per cent)	Frequency (per cent)	Frequency (per cent)	Frequency (per cent)
B. c. maxima						
Immatures	35	45.7	25.7	20.0	8.6	0.0
Yearlings	21	38.1	23.8	19.0	14.3	4.8
Adults	23	17.4	30.4	26.1	17.4	8.7
B. c. interior						
Immatures[1]	627	95.4	—	—	—	—
Immatures	24	79.2	12.5	8.3	0.0	0.0
Yearlings	17	52.9	35.3	11.8	0.0	0.0
Adults	36	64.0	19.4	16.6	0.0	0.0
Yearlings and adults[2]	404	83.7	—	—	—	—

1. From Elder (1946). 2. *Ibid.*

The per cent of *interior* recorded in Table 14 as having a neck ring is higher than that recorded by Elder (1946). The difference is probably due to Elder's more rigorous requirements for identifying a ring. In the present data, geese which had a trace or scattering of white feathers around the neck are included in the first category — those having a barely discernible ring. Data for both races show a greater prevalence of white neck rings in the older age classes.

COLORATION OF THE DOWNY YOUNG

The coloration of the downy young of the races of Canada geese varies widely and is, therefore, useful in determining racial affinities of a population (Hellmayr and Conover, 1948:300). Delacour (1954:163) has described the downy young of *maxima* from Minnesota and North Dakota as follows: "much as in *canadensis*, but larger and more golden, less brown, above; basal line and circles around the eyes very pale, hardly indicated. . . ." The downy young of *moffitti* are reported by Delacour to be like *maxima* in color; in contrast, those of *canadensis* (p. 158) "have the forehead, sides of head, skin, throat and underparts clear yellow, paler on the throat and belly, passing to olive-green on the flanks, lores, a small patch around the eyes, crown, hind neck and upperparts olive-green with yellow patches on wings and back."

Regarding the downy young of *interior*, Delacour commented, "much as in *canadensis*, perhaps a little darker." However, goslings of *interior* that I have collected on Akimiski Island and my photographs of *interior* from northern Ontario reveal that the downies of *interior* are altogether different from Delacour's description of *canadensis*, the color of the back being markedly darker — basically a drab brown in color with only a hint of olive green tones on the longer down. The under parts are distinctly grayish and the crown of the head is dark brown (Fig. 21).

In addition to the American Museum specimens described by Delacour, I have examined two other specimens of *maxima* from North Dakota (CNHM). The most notable characteristic of all these downies is the dark gold or golden tan color of the back and general pale buff color of the forehead, sides of head, and underparts. It is apparent from a reproduction of a color photograph I have seen of a pair of adults with two sets of downy young, presumed to have been taken at Rennie, Manitoba, that the downies of the geese breeding in this area are identical in coloration. Like the North Dakota specimen at hand, the Rennie young clearly lack a dark ring around the eyes. Three downy young of a series of goslings collected at the Seney National Wildlife Refuge (MMZ) are basically similar to the North Dakota *maxima* but are slightly darker (Fig. 21).

FIELD IDENTIFICATION

An experienced observer should have little difficulty identifying *maxima* either at a distance (with the aid of field glasses) or close at hand if he is familiar with any of the other large races. A summary of useful field characters may substitute for a more formal review of the physical characteristics of *maxima*.

SIZE AND BODY PROPORTIONS — One-fourth to one-third larger than the other large races; neck proportionately the longest of all races of *canadensis;* bill massive, usually with a blunt rather than tapered tip; wing span of adult males 6 feet to 6½ feet; scutellation of tarsi prominent; feet extremely large.

COLORATION — Breast color much lighter than *interior,* especially the males of stocks from Manitoba. When flocks face the sun, or are seen flying overhead, breast appears pearly gray or almost white (Fig. 26). Differs from *canadensis* by lacking a whitish mantle over the back just below the white stocking, but superficially similar in color to *moffitti.* Differs from *interior* by having a high frequency of neck rings which are pure white and often wide. (Neck rings are present in *interior* but are usually dusky white and narrow.) A white spot or bar across the forehead or a white spot above each eye occurs in a few individuals of most populations of *maxima;* they are absent or rare in the other large races. The cheek patches are extensive, reaching the crown of the head in some individuals. The recognition of *maxima* is virtually certain if the top of the cheek patches of some individuals in a flock have a small hook or projection extending posteriorly.

FLIGHT — Wing beat relatively slow; shallowness of stroke distinctive. Local flights often made at low elevation. The long wings and neck of this race are conspicuous in flight (Figs. 26 and 27).

VOICE — Usually silent in flight, calling only infrequently, and its voice is of lower pitch than that of the other races, especially in birds of the largest sized stocks.

NOTES TO CHAPTER 2

[1] These clinal relationships will be discussed in detail in a forthcoming paper on the taxonomy and evolution of the races of Canada geese.

[2] "Foxing," the tendency of brown feather pigments to take on a reddish cast after skins have been held for many years in museum cases, was not a factor as, with only one exception, critical comparisons and descriptions were made from frozen specimens or skins that had been collected two years or less prior to study.

[3] The geographical origin of most of the large dark-colored stocks that are captive or are now wild descendents from transplanted former decoy flocks is unknown; however, I believe most are the progeny of birds originally captured in the wild in the more eastern and southern portions of the original prairie range. That many of the progenitors of decoy stocks that subsequently became widely distributed across the continent came from Iowa, is indicated by Huntington's (1910:137) extract of an article by Warren R. Leach:

> It was some time in the seventies that my brother called my attention to an advertisement of a party in Fort Dodge, Ia., in one of the sporting magazines who offered Canada wild geese for sale. Geese were then nesting plentiful in parts of that State, and those offered for sale were goslings captured from adjacent sloughs.

[4] Wild, migrant stocks of giant Canada geese studied at the Squaw Creek National Wildlife Refuge, Missouri, in December 1964, were found to be decidedly more tannish in coloration than the flocks wintering at Rochester, Minnesota, and in Rock County, Wisconsin. They were, however, found to be similar in size.

Breeding Range 3

THE BREEDING range of *maxima*, as delineated by Delacour (1951) on the basis of Moffitt's earlier unpublished studies, was reported to have extended from the Dakotas and Minnesota south to Kansas, Kentucky, Tennessee, and Arkansas. This range, I assume, was based on the specimens available in museums, the writings of Mershon, and the reports of early naturalists and explorers, but it ignored early records of breeding Canada geese in peripheral areas. Like previous versions of the distributions of the races, it was not based on a realistic consideration of the ecological, physiographic, geological, and climatic factors involved. The acumen with which evolution and distribution have been studied in so many other species of birds has not been demonstrated in speciation studies of Canada geese; yet it will be shown that the distribution of the races of Canada geese displays with classic clarity the basic evolutionary principles and environmental relationships (Hanson, *unpublished*).[1]

In the past, large Canada geese nested in the states and provinces which are peripheral to the range given by Delacour: portions of Colorado, Wyoming, and Montana east of the Rockies; the lake states of Wisconsin, Illinois, Indiana, and Michigan; and in Canada, the prairie provinces; possibly the southern portion of extreme western Ontario; and southern Ontario. Nesting still occurs in most of these areas. In Map i the approximate range occupied by *maxima* in the past is shown. This entire area, with only a few exceptions, lies within the prairie biome or its adjacent ecotones.

The limits of the range which have been occupied by the giant Canada goose are easily described in terms of geological, physiographic, and ecological boundaries (Map ii).[2] With minor exceptions, its primary range has been confined to the tall-grass and mixed prairie areas with

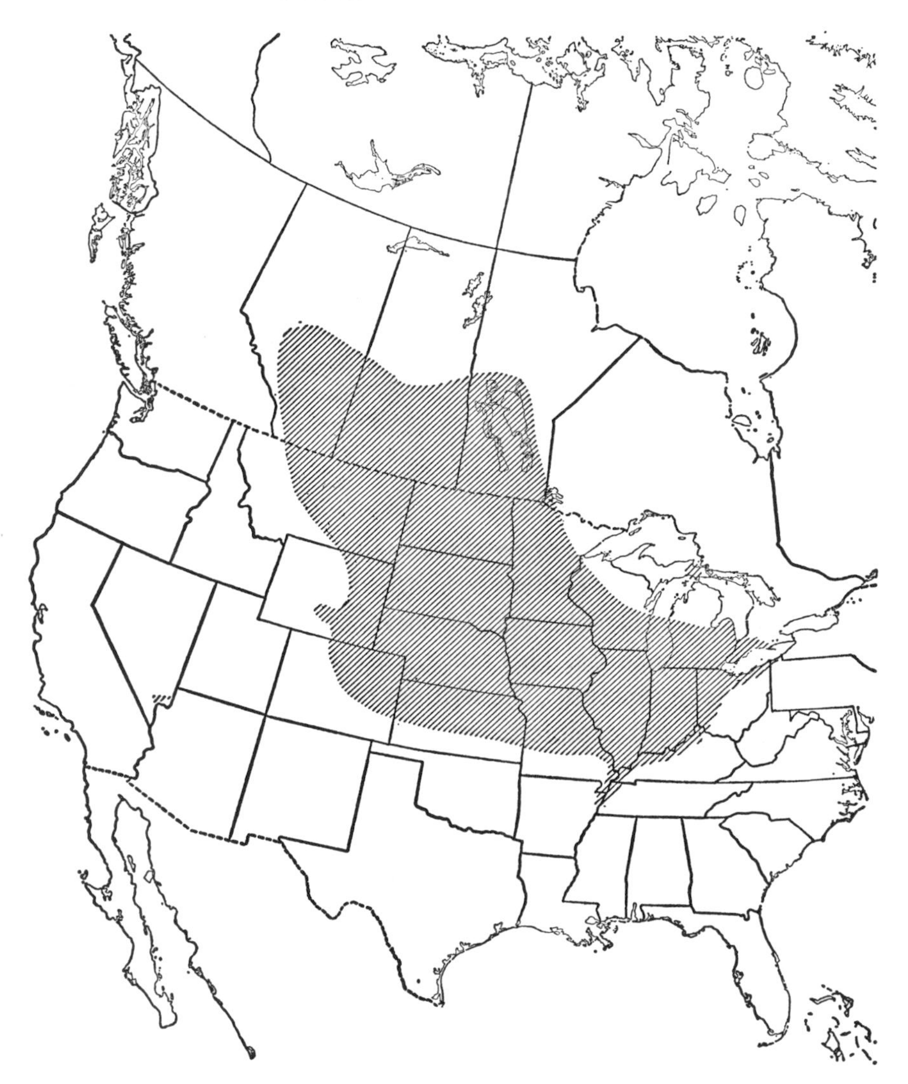

i. Approximate breeding range of the giant Canada goose prior to settlement.

their lake and marsh complexes that lie north of the limits of the Wisconsin Glaciation which, in Nebraska and the Dakotas, closely follows the Missouri River. Some nesting occurred in country west of the Missouri River — the Great Plains Province as defined by Fenneman (1931) — but, because of its more rugged character and fewer water areas, it probably held very much lower densities of nesting geese in the past than the eastern prairies. Included in the range of *maxima* are the prairie portions of the Canadian provinces and the adjacent areas of aspen ecotone. In the western sectors of its range, heavy coniferous

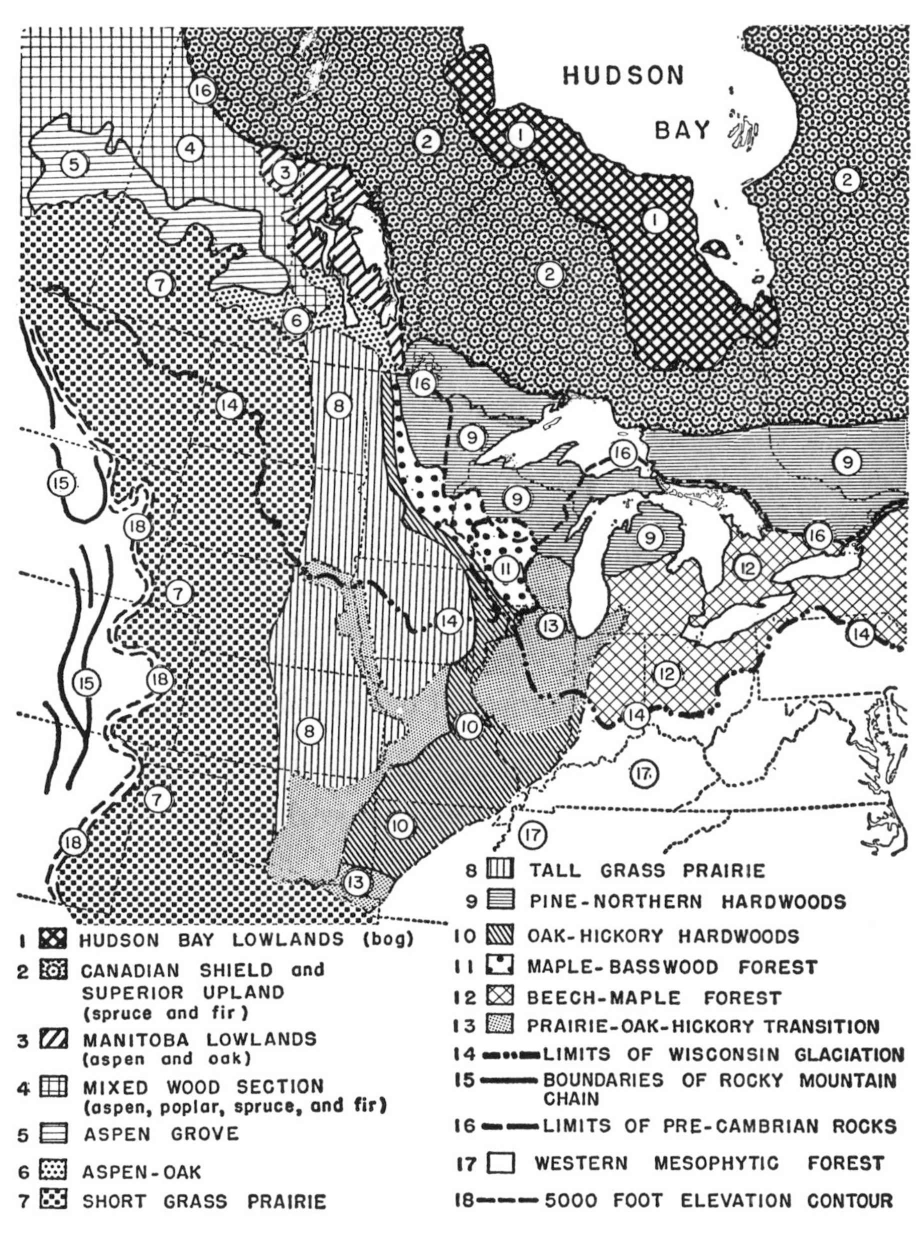

ii. Limits of various physiographic, geological, and ecological areas that determined the range of the giant Canada goose in prehistoric times.

forest limited its distribution northwards; in the eastern sectors, the south portions of the rocky, forest-covered, and relatively sterile Canadian Shield formed the northern boundary of its range. In the United States, contiguous elements of the Canadian Shield, the Superior Uplands (the limits of which are not well demarcated because of thick, overlying layers of glacial drift) have also imposed northward limits to its range.

South and west of the Shield in Manitoba, the breeding range includes the forest-grassland transition zone (Map ii) in which aspen (*Populus tremuloides*) is the predominant tree — the interlake region and the Saskatchewan River delta country, which Rowe (1959) has classified as the Manitoba Lowlands. Here the land is poorly drained and characterized by "forest patches of black spruce (*Picea mariana*) and tamarack (*Larix larcina*), with intervening swamps and meadows" (Map ii). Between the latter zone and the Hudson Bay Lowlands, which contain the breeding grounds of the Hudson Bay Canada geese (*interior*), the Canadian Shield forms a belt of inhospitable terrain approximately 200 to 400 miles wide which acts as an effective barrier isolating the two breeding populations from one another.[3] To the south, extensive forests undoubtedly limited the range of *maxima* eastward into Indiana and Michigan, although pockets of prairie, large marshes, and lake areas held nesting populations. Similarly, lake and marsh complexes in the oak-hickory and prairie transition region of Illinois and southern Wisconsin afforded breeding areas.

The range of *maxima* was, in part, presaged by the glacial lake systems that were formed with the retreat of the ice sheet of the Wisconsin Glaciation. In historical times, the densest populations of *maxima* appear to have been associated with the areas formerly occupied by these lakes — particularly the postglacial extensions of Lake Erie, Lake Huron, Lake Michigan (Lakes Maumee, Alma, Chicago, Wauponsee, Watseka, Pontiac, Ottawa, Oshkosh) and possibly Lake Wisconsin. On the Great Plains, nesting areas of primary importance were created by Lake Agassiz, Lake Dakota, former extensions of Devils Lake, Lake Souris, Lake Regina (which occupied the basin of the South Saskatchewan River) and possibly the former glacial lake areas in the basin of the upper Missouri River in eastern Montana (Flint, 1945, 1957, and Flint et al., 1959).

Prior to settlement, the giant Canada goose must have nested around much of the periphery of the present-day Great Lakes system, depending on the distribution of suitable sloughs and marshes. I would exclude from this generalization the rocky shorelines of the north shore of Lake Superior and the north shore of Lake Huron. Probably most or all of Lake Ontario was east of the original breeding range of this race.

The ecological associations of the prairie biome have one important factor in common — highly calcareous top soils or subsoils. According to

Rowe (1959:21), "the underlying bedrock of the Manitoba Lowlands is Palaezoic limestone, and the overlying beds are mainly lacustrine clays deposited in glacial Lake Agassiz or modified chernozems. . . . The influence of the limestone parent materials can be seen in the soils which tend toward rendzinas and high-lime meadow and peat profiles." The prairie soils of the tall-grass prairie and the chernozem soils of the short-grass or mixed prairie areas to the south have a high lime content, particularly the latter, as exemplified by the alkali conditions of its lakes. Early records of giant Canada geese breeding outside these soil types are generally associated with flood plain areas of the major river systems, the soils of which were enriched by annual flooding — for example, the lower Missouri, Ohio, and the upper Mississippi rivers.

The past and present distribution of breeding populations of *maxima* in the Canadian provinces and the United States is detailed below.

Canada

ONTARIO — Records of Canada geese nesting in the southern portions of the western edge of Ontario probably represent penetrations by *maxima* into the periphery of the Canadian Shield country. In 1961, personnel of the Ontario Department of Lands and Forests observed geese on the portage between Gammon and Hammerhead lakes, approximately long 94° 50′ W, lat 51° 00′ N.

Early in this study it was concluded that Lake St. Clair must have been an important breeding area for giant Canada geese in primitive times. It was, therefore, most satisfying when I was examining skins in the Royal Ontario Museum in August, 1963, to find a specimen, judged from various age and sex criteria (Hanson, 1962) to be an adult female, that had been collected on April 25, 1884, at Lake St. Clair (Table 7). The date of collection establishes this specimen as a resident of the area and a potential nester although it lacks a brood patch.

Harry G. Lumsden has called to my attention the following remarks by M. de Lemothe Cadillac, the commandant of the French Detroit settlement, in a description of the area on October 8, 1701, which lend support to the above conclusions (Lajeunesse, 1960, p. 19):

> There are such large numbers of swans that the rushes among which they are massed might be taken for lilies. The gabbling goose, the duck, the teal and the bustard are so common there that, in order to satisfy you of it, I will only make use of the expression of one of the savages of whom I asked before I got there whether there was much game there, "there is so much," he told me, "that it moves aside to allow the boat to pass."

Mr. Lumsden has also called my attention to the record of a humerus bone of a Canada goose recovered from an Indian occupation site at Port Maitland, Ontario, near the east end of Lake Erie. The age of the site is approximately 1,300 years. This humerus measures 201 mm in length which would qualify it as being from a large specimen of *maxima* (Table 31). A photograph of this bone, misidentified as being from an eagle, is shown in Ritchie (1944: Plate 83).

MANITOBA — Inclusion of parts of Canada adjacent to the range of *maxima* in the United States apparently was not documented by Moffitt in his unpublished notes and was not included in the range given by Delacour (1951 and 1954). The reason for this omission seems clear: all museum skins of *maxima* available at that time had been collected in the United States (Table 7) and, as the race had been presumed extinct, the scattered populations present today in southern, central, and northwestern Manitoba and adjacent Saskatchewan were presumed to be *moffitti* (Delacour, 1954:165), and more recently *interior* (Klopman, 1958). The Indians have sometimes been more acute in distinguishing the various races of Canada goose than the taxonomist. At Lake St. Martin an Indian reported to P. A. Taverner (Shortt and Waller, 1937:10) that four kinds of Canada goose occurred in this area and that they could be distinguished only by size and minor characters. Regarding the largest of the four described, Taverner remarked, "seems more or less traditional. It seems to be an immense bird and is so rare that it is known only by report. It is probably mythical." The local Indians were, of course, referring to the giant Canada goose and had no doubt stressed the huge size attained by some adult males. However, Canada geese of exceptional weight continue to be reported in Manitoba from time to time (see Table 1). In the *Dufferin Leader*, November 19, 1962, there is the following account: "While out shooting with Marshall Harrison Wednesday, November 11, near Mariapolis, C. W. Orchard bagged what is believed to be one of the largest wild geese shot in that area in recent years. The Canada Goose tipped the scales in both Graysville stores at 17¼ pounds. Mr. Orchard says this gander was almost twice the size of any of the other four he shot that day" (*newspaper clipping*, courtesy of Alex J. Reeve).

In the summer of 1964, I was told of a goose weighing between 20 and 23 pounds that was shot close to Lake Winnipegosis in the late 1940's (Ivan Bratke, *personal communication*).

The present study has fully established that the Canada geese breeding in Manitoba south and west of the Canadian Shield are *maxima*. (*See* various tables for weight and size data and Figures 1, 28, and 29 for body proportions and coloration.)

The past and present distribution of Canada geese in Manitoba south of the Canadian Shield is unusually well documented. Macoun (1883: 364–65) was one of the first to comment on its status: "The Canada goose still breeds around the larger lakes, and on islands in them. So late as June, 1879, I have seen the Canada Goose and a flock of young ones on the Assiniboine where the town of Brandon now stands, and at present they still breed on the river, above Fort Ellice."

Thompson (1891:486–87) extensively summarized early records of Canada geese in the province. The following breeding records are presumed to pertain to *maxima:* Winnipeg — summer resident, abundant, a few breed; Red River valley — a few breed; breeding on Swamp Island, Lake Winnipeg; Ossowa — common, breeding 1885; Portage la Prairie — a few breed in the marshes near Lake Manitoba; Shell River — odd pairs breed near here (about 1885); Qu Appelle — common summer resident; Carberry — rarely breeding south of Souris River; West Shoal Lake, breeding.

In 1900, Preble (1902:90) was shown an island in the northern part of Lake Winnipeg where considerable numbers of Canada geese nested.

Bent (1925:210) observed Canada geese in west-central Manitoba. "While cruising on Lake Winnipegosis on June 18, 1913, we came upon a family party fully 5 miles from shore and evidently swimming across the lake! The two old birds when hard pressed finally took wing and flew away, leaving the three half-grown young to their fate." Arthur S. Hawkins (*personal communication*) stated that in recent years he has observed Canada geese along the west side of Dawson Bay, Lake Winnipegosis.

On July 6, 1920, Rowan (1922:226) flushed two Canada geese from a small marsh near Indian Bay on Lake-of-the-Woods. The 40-mile stretch of muskeg that Rowan reported from that area may be presumed to hold nesting Canada geese at present.

Shortt and Waller (1937:9) reported on the status of the Canada goose in the Lake St. Martin region: "It is not a common summer resident, but Indians have found nests of this species in the muskeg about the mouth of the Dauphin river."

In 1953, Godfrey (1953:6) made an ornithological survey of Manitoba reporting, in addition to his own observations, records in the National Museum of Canada made by P. A. Taverner:

> Breeds in the Boggy Creek marshes near Whitemouth (eggs and downy young taken by Robert Latta of Whitemouth for propagation purposes). Taverner and Lloyd in 1921 saw eight at Clandeboye Bay on July 8, and eight at Oak Point, July 11. On Lake Winnipegosis marshes near Overflowing River, the writer saw one on July 24, fourteen on July 26, and forty-two on

August 2, 1951; also a photograph of an adult and four downies taken by Leo LePine on June 11, 1951, on Overflowing River was examined. Occasional flying flocks of about six birds noted by Taverner near The Pas, June 2 to 20, 1937. At Herchmer an adult and six goslings were noted on June 29, 1936, by Taverner. [This latter record undoubtedly refers to *interior*.] No breeding specimens are available for subspecific determination.

Bird's (1961:72–73) outstanding study of the ecology of the aspen parklands of Canada has provided a number of valuable historical records not available elsewhere:

> The Canada goose, *Branta canadensis* (L.) is the only goose that nested in the parkland. It apparently nested generally throughout the area but there is little exact information on its nesting locations. Stuart Criddle (naturalist and old-time resident, Treesbank, Man. In litt. 1955) stated that a few pairs nested along the Assiniboine River near Treesbank, Manitoba, in the 1880's and that some nested in the Douglas marsh, ten miles to the north, until the early part of this century. Hales (1927) stated that he shot a Canada goose out of a flock nesting in Oak Lake, Manitoba, in 1901, and that geese were then nesting in only a few scattered localities. According to Lawrence (1925), A. G. Vidal of West Shoal Lake, Manitoba, reported that 20 Canada geese had nested there in 1925 and that these were the first geese nesting there for about 30 years. This was just a year following the creation of the Shoal Lake sanctuary in February, 1924. . . . A few pairs still nest in more isolated marshes, e.g., Dog Lake near Ashern, Manitoba, where a sizeable colony is still found; but the goose has been practically exterminated as a breeding bird in the aspen parkland. It is easily raised in captivity, and some captive birds have been released and are once again nesting in the old haunts in a few protected marshes.

The Saskatchewan River Delta as a whole may contain a fair population of giant Canada geese but, in the opinion of Jack L. Howard of the Manitoba Department of Mines and Resources, the actual nesting density is light. An aerial search of this area made by Harry G. Lumsden and me in July, 1962, seemingly supported this conclusion; between The Pas and Cedar Lake we saw only one flock of adults and young, totaling about 75. However, unless Canada geese are concentrated on rivers and lakes where they can be easily seen, aerial search can be an ineffective way of finding them on their breeding areas. The delta of the Saskatchewan River is a vast and fairly inaccessible country; it would, therefore, seem very desirable to record in detail some of the observations of Canada geese that have been made by various ground parties in the general area in recent years.

On May 30, 1946, Arthur S. Hawkins and a party went ashore at

West Mossy Portage, between Cedar Lake and Lake Winnipegosis. They immediately saw two pairs, whose behavior indicated they had nests nearby, and a third male was heard in the distance. Later, they went ashore near East Mossy Portage and flushed a pair and, around the next point, a second pair. In the next three bays they flushed a total of six pairs, all in an estimated 6 miles of shoreline.

On July 24, 1947, a party engaged in river basin studies surveyed 179 shoreline miles of Cedar Lake. They observed a total of 68 adults and 36 broods, the latter averaging 3.7 young. They also observed 2 broods on the lower Saskatchewan Delta and 11 broods on Lake Winnipegosis, which averaged 4.3 young; 4 broods with 10 young on Waterhen Lake; 26 broods which averaged 4.4 young on Lake Manitoba; and on Lake St. Martin, 3 broods with 14 young.

Jack L. Howard observed broods of Canada geese on the following lakes: Lake 52 (just south of Mawdesly Lake), Lake 9 and Lake 10 (just south of Landry Lake), Lake 4 (west of Landry Lake), Kelsey Lake (south shore), Little Fish Lake; Long Grass Lake; and Head River Lake. He had reports of broods on several lakes south of Moose Lake Settlement and on Connolly Lake. Charles H. Lacey of Ducks Unlimited reported goose broods on Lake 6 (north of Watseekwatapi Lake), Lake 15, Lake 19, Elm Creek (west of Saskeram Lake), and on the floodwaters above the Prairie Farms Rehabilitation Act diversion dam on the Pasquia River.

Much of the vast delta of the Saskatchewan River will soon be flooded by a dam being constructed at Grand Rapids. The subsequent displacement of the nesting population of Canada geese in this area should be studied.

Nesting of Canada geese is also known from the Setting Lake and the Sipiwesk Lake areas on the Precambrian Shield 75 miles north of Lake Winnipeg (Arthur S. Hawkins, *personal communication*). Broods have also been seen on Gormley Lake and Restines Lake north of Lake Winnipeg and close to the Hudson Bay railway line (William Leitch, *personal communication*).

The large shallow lakes of the interlake area of the Manitoba Lowlands between lat 54° 40′ N and Sturgeon Bay on Lake Winnipeg form the most extensive block of habitat for Canada geese in the province. Breeding geese are known to nest in the area bounded by Kawinaw and Katimik lakes in the north and Chitek Lake in the south (Arthur S. Hawkins, *personal communication*). On the latter lake, Harry G. Lumsden and I observed two pairs in July, 1962, and collected one pair for identification of subspecies, which proved to be *maxima*. Charles H.

Lacey has stated (*personal communication,* December 7, 1962) that many of the lakes of this area support one or two pairs. Although Harry Lumsden and I saw few geese in this area, it should be pointed out that it is difficult to see geese (except on water areas) in making aerial surveys of bush and muskeg country as compared to the ease with which they may be observed in Arctic areas.

Lacey also reported that 20 to 50 pairs breed in Hecla Island in Lake Winnipeg. Large numbers of goslings have also been reported along the Moorehen River which drains into Lake St. George southeast of Sturgeon Bay. According to a resident trapper familiar with the country, a minimum of 50 pairs nest in the area lying between the village of Hodgson and Washow Bay, Lake Winnipeg (Eugene F. Bossenmaier, *letter,* September 18, 1963).

Lacey reported that as many as six pairs nest in the vicinity of Rivers, Manitoba; at Whitemouth Lake, five broods were reported in 1962, and at Fort Whyte, Manitoba, a suburb of Winnipeg, there are 22 adult geese in a flock breeding under semicaptive conditions. These geese were originally obtained from a captive flock near Morden, near the Manitoba–North Dakota border. I found these Canada geese to be very large and possessing the typical characters of *maxima* (Fig. 28).

The present-day population of giant Canada geese in Manitoba includes four accessible nesting concentrations: Dog Lake, approximately 400; the Delta marshes, 300; the East Meadow Ranch (near Oak Point), 400 (Fig. 29); and the immediate area of the Alf Hole Provincial Wildlife Refuge near Rennie, where there were 12 breeding pairs in 1963. Both the East Meadow Ranch and the provincial refuge are also fall staging areas. By late September or early October, the East Meadow Ranch may hold 5,000 to 6,000 giant Canadas and the provincial refuge, approximately 500. The nesting population in the Delta marshes was established from stock originating near Rennie, and the population at the East Meadow Ranch was derived from 11 birds taken from the Delta flock.

According to information obtained by George C. Arthur and me in the summer of 1962, a definite increase of scattered nestings in small marshes and sloughs has occurred in recent years in the area east of Lake Manitoba between Oak Point and Dog Lake. These nestings are a source of considerable pride to farmers on whose land they occur. Partially as a result of the protection received from farmers, the Manitoba populations have been steadily increasing.

In 1961, several pairs were reported to have nested on an island in the narrows of Lake Manitoba west of Dog Lake. In 1964, seven pairs nested on a small island at Bluff Harbor on Lake Manitoba near Reykjavik (Fig. 47).

SASKATCHEWAN — Weights alone of large Canada geese from Saskatchewan may be considered almost conclusive evidence that the race breeding in the southern half of the province is *maxima*. The first information obtained indicating this is a letter in the Mershon files. William B. Mershon, Jr. to William B. Mershon, Sr., June 9, 1924:

> Here is some more information on the big goose. Last fall Bert Horseman, the Customs Inspector at Moose Jaw, killed a young goose near Johnstone Lake that weighed 13½#. Cecil Cull of Moose Jaw Cold Storage tells me that two or three years ago a goose was brought in weighing 16#. He thinks the latter was shot also near Johnstone Lake. He doesn't remember the exact date.

Wherever Canada geese are hunted, there are almost certain to be local contests for the largest goose shot each season. Goose hunting areas in Saskatchewan are not exceptional in this respect. A news clipping in the *Saskatoon Star-Phoenix* (December, 1962) furnished by Tom Sterling provides some current information on the size of Canada geese shot in the province. At Eston, a 12 lb. 5 oz-bird won the local 1962 competition; at Invermay, the heaviest goose shot weighed a half ounce under 14 pounds and the runner-up, 13 lb. 14 oz.

Albert J. Boswall, a game bird breeder on Prince Edward Island, recently reported (*letter* to H. Albert Hochbaum, January 10, 1963) that he had several captive Canada geese, originally obtained near the village of Hudson Bay, Saskatchewan, which weighed over 15 pounds when in good condition. According to Harvey W. Miller, a goose weighing 16 to 18 pounds was shot at Foam Lake, Saskatchewan, in 1961.

In regard to maximum weights attained by Canada geese in the prairie country of Saskatchewan, Alex Dzubin (*letter*, January 3, 1963) has commented:

> Perhaps there are local populations of *maxima* scattered across the prairies as reports of 14 and 15 pound birds are met with every fall. I haven't seen or measured any of these.
>
> I talked to an old goose hunter from Govan, Saskatchewan, on the northeast side of Lost Mountain Lake who claimed to have regularly shot 16–20 pound geese prior to 1928. Since this time he says very few really large Canadas are found in that district.

Weights of geese shot by hunters are sometimes exaggerated, and I would put myself in the front ranks of those who have doubted reports of geese of exceptional weight. Now, however, with the advantage of the hindsight which the present findings afford, the following additional comments of Dzubin (*op. cit.*) must be appraised in a new light:

> Every small town in western Saskatchewan has a story (or fable) about the 18 pounder that was killed by some one or other away back in 1932 or 1945. The veracity of these stories is questionable even though good numbers of reliable witnesses swear to the weights of big geese. I believe though that a large Canada of this type is to be remembered for a long time and is an event to be marked and talked about for many years by local people. Few hunters brag about their 11–12 pounders [They would at Horseshoe Lake, Illinois!] but everyone remembers one bigger.

Tom Sterling of Ducks Unlimited has provided a review of the current status of Canada goose populations in Saskatchewan (*letter*, December 11, 1962):

> Canada geese formerly bred throughout eastern Sask., but not long after Settlement (1900–1910) they survived only on a few of the larger lakes and marshes of the parklands (Quill Lakes, Kinistino, Tisdale and Yorkton areas), in large marshes around the Pasquia Hills and in the Saskatchewan River delta. Since the 1950's there has been a marked increase in numbers and the total breeding population would be somewhere between 200–300 pairs. Waterhen marsh (near Kinistino) drained in 1920 and reflooded in 1939 had approximately 60 broods in 1962. . . . There are a *few* geese again breeding in the wild in southeastern Sask., believed to be spreading from introduced flocks at Abernethy and Regina. I believe both these flocks were started from eggs collected in the Quill Lakes area. The Regina (Wascana) flock now approximates 50 breeding pairs, and population pressures are forcing a natural spread. . . .
>
> Southwestern Sask., South of the South Sask. River and west of the 3rd meridian. There were Canada geese breeding in this area at time of first settlement (1880–1910) but numbers are unknown. By 1930 the population was very low and during the drought of the 30's may have been completely wiped out except for those on the Sask. River. Since the 30's the population has built up to an estimated 250 pairs by 1962 (actual count of 147 broods in 1959). Indications are this repopulation was due to Montana breeders spreading northward. . . .
>
> Western Sask., South Sask. River north to the North Sask. River. This area had local breeders throughout when first settled (1900–1915), even through the pothole areas of the Missouri Cotteau. Settlement and the drought of the 30's eliminated them except for those on or moving to the South Sask. River. There are now a few pairs scattered throughout this area from the South Saskatchewan to Manito Lake. We know of no birds breeding in western Sask. north of the North Saskatchewan River. An aerial census conducted in April, 1961, indicated 80 pair distributed along the

> South Sask. River from the Alta. border to the Elbow. There are probably no more than 20 pair in all the rest of this part of western Sask. . . . We have introduced flightless young geese to a small section of this area. The stock was obtained from the Bowdoin Refuge in Montana. Releases were as follows: 1960 — 86; 1961 — 49; 1962 — 69. . . . Pairs were noted on the releasal area this spring but attempts to locate nests or early broods were unsuccessful. One banded pair of geese was seen with brood obviously just on the wing, in early August. One of the adults was collected and proved to be a male from the 1960 release.

Dzubin (*letter,* January 3, 1963) has furnished additional detail on the distribution of large Canadas in Saskatchewan:

> Large type Canadas are found breeding into the grassland-parkland ecotone from Manito Lake west of Battleford along the north Saskatchewan River to a point north of Saskatoon. I've seen apparently breeding pairs of Canadas 20 miles north of Wakaw and then eastward to Melfort and southeastward to Kamsack on the Manitoba border. My impression has been that the birds [Canadas] prefer grassland or open type habitat rather than forested areas. . . .
>
> In the Kindersley district most permanent lakes have 1 or 2 pairs which yearly nest there and raise broods. . . . In Saskatchewan I would estimate 125–150 pairs breeding between the North and South Saskatchewan Rivers including rivers themselves. . . . the S. Saskatchewan River where it enters the western edge of the province to a point immediately north of Swift Current . . . containing the heaviest concentration of Canadas in the central regions of Saskatchewan.

James L. Nelson has provided further insight into the status of Canada geese nesting on the Saskatchewan prairies (*letter,* January 3, 1963):

> Alex Dzubin . . . has suggested that I write you regarding the surveys the Saskatchewan Wildlife Branch has conducted in the southwestern part of the province. These brood surveys began in 1960 and our data was combined with data from similar surveys by Ducks Unlimited to determine the approximate breeding population and population trend in the southwest. The general area involved is bordered by Alberta on the west, Montana to the south, the South Saskatchewan River on the north, and Highway #4 on the east.
>
> Our most complete brood count was made in 1960 when we observed 203 broods in this area (including the South Sask. river) and had reliable reports of a few others. In all there could possibly have been 250 broods. Since 1960 the water areas have continued to deteriorate throughout much of the census area and brood numbers have declined slightly. Since very

few of the water areas are censused prior to the late June brood count we know very little about nesting success. It appears that the non-breeders and unsuccessful breeders move off to more preferred molting areas prior to the brood counts. This definitely occurred on the South Sask. River where a more detailed study was made in 1962. In this case no adults were seen on June 29th that weren't associated with broods whereas on May 25th geese in flocks composed about half the total adult population. [These nonbreeding geese molt in the Arctic. *See* Chapter 4, "Migration."]

Nesting success on the South Sask. River sample of 38 known-fate nests was 42.1%. I believe that this is considerably lower than nesting success on the other areas in the southwest but can't substantiate it with any data. Assuming an overall nesting success of 60% however, you could place the breeding population at roughly 400 pairs which I don't think would be too far out of line.

In general I think the numbers of geese in the southwest and the northeast would be comparable, possibly even more in the northeast. Widely scattered pairs also nest in the area between the North and South Saskatchewan rivers, but the density is pretty thin compared to the other areas mentioned.

ALBERTA — Until recently, the lack of specimens from the prairie provinces of Canada in major eastern museums contributed to the belief that the race *maxima* was extinct and never had been a breeding bird in this vast area. However, information available leaves little doubt that the range of *maxima* does extend across southern Alberta to the foothills of the Rocky Mountains. Weights of Canada geese shot in Alberta alone suffice to indicate the occurrence of *maxima* in this province.

W. F. H. Mason, Northern Alberta Game and Fish Protective League, to William B. Mershon, June 30, 1924:

From information I have got myself I find that Dr. F. J. Folinsbee of this city [Edmonton, Alta.] shot a goose, who he states was the leader of the flock and after getting the head and neck off, the cold storage weight was sixteen and three quarter pounds.

Dr. R. A. Rooney last year shot a goose just tipping the scales at sixteen pounds.

Bert Laird and Wm. Pigeon of Wainwright, who are great hunters, inform me that they have shot geese weighing sixteen pounds each, but have not got any during the last few years. These men state that these geese came to the field alone and did not mix with any of the others but they were unable to inform as to their being any different from the true Canada or Hutchins goose. [The writer is undoubtedly referring to one of the lesser Canada geese, *B. c. parvipes*.]

Regarding weights of Canada geese in Alberta, R. Webb, a wildlife biologist for the Alberta Department of Lands and Forests, (*letter*, January 30, 1963) wrote:

> I doubt very much the validity of your 15–16 lb. observation [I quoted from the Nicholson article in *Rod and Gun* in a memorandum sent to Canadian collaborators re. the Calgary birds; see below.] I have weighed no adult males (or any others) over 13 pounds. A very few up to 14 lb. have been reported in the last few years but none verified by responsible people.

The biologist familiar only with a population of large Canadas containing adult males weighing 12 to 14 pounds may not consider such geese as being of exceptional size, but in comparison with adult males in populations of *interior* (see Table 2), *moffitti*, or *canadensis*, they are indeed large birds. For example, in the mountain valleys of adjacent British Columbia where *B. c. moffitti* breeds, a 10-pound Canada goose is reported to be a "whopper" (D. J. Robinson, *letter*, January 21, 1963). Note in Table 3 that the weights of five specimens of *moffitti* sent to me from Utah are similar to *interior*.

A photograph and letter to the editor of *Rod and Gun in Canada* in the 1940's describing the Canada geese which winter at the Inglewood Bird Sanctuary, East Calgary, virtually confirmed the fact that *maxima* is the Canada goose breeding in the southern portions of Alberta east of the Rockies. As could be seen in the photograph, the geese were very large, with long bodies, long necks, and light color — all characteristic of the race. Close inspection of the photograph revealed that most of the geese had white neck rings and large bills. One large individual, a gander, had a prominent white blaze mark across the forehead. An iron stain on the breast is said to characterize this wintering population. The refuge manager at that time estimated that the geese wintering there weighed up to 15 or 16 pounds. Support for this estimate was found in a letter of the late William Rowan, the noted ornithologist of the University of Edmonton, to W. B. Mershon, January 14, 1926: "At Tofield, Alberta, there was shot by a friend of mine a Canada goose, that, as far as I remember weighed about 15 lb. This was in the fall of 1924. . . . This bird was mounted on account of its exceptional size, and is now at Tofield."

Subsequent to the realization that the flock wintering at Calgary was *maxima*, Harry G. Lumsden sent me photographs of the geese that nest at the Inglewood Sanctuary (Fig. 30). Careful inspection of these photographs revealed that these geese had all the characteristics of *maxima* and were seemingly identical to the population nesting in Manitoba.

An observation made by Robert H. Wheeler, a United States Game Management Agent (*letter*, April 2, 1963) in central Alberta in the summer of 1959, leaves little doubt that the Canada geese he observed were *maxima:*

> Near the end of July while scouting for duck trapping sites by boat on Whitford Lake, immediately southeast of Andrew, Alta., I observed nine very large, dusky colored Canada geese. At this time, I was stationed in the Southern District of Illinois and was very familiar with Canada geese [*B. c. interior*]. These geese, viewed from one hundred yards, appeared to be extremely large, and when they flew, their movements appeared to be more ponderous and laborious than those of any Canada geese which I had previously observed. [See Flight under "Field Identification."] I know nothing of the coloration of the Giant Canada geese, but these were much darker than the average Canada goose. I was quite puzzled at the time, as I knew nothing of a race of Giant Canada geese, and have thought of them many times since.

Clippings from newspapers have been invaluable in documenting the range of the giant Canada goose. The last to be received (courtesy of W. G. Leitch) is from the *Winnipeg Free Press* of September 21, 1963. I believe it removes the last vestige of doubt that the giant Canada breeding range extends across the prairies of southern Alberta. Under a photograph of an obviously very large Canada goose the legend reads: "HUGE HONKER – Honkers are plump and plentiful this season in central Alberta's grain country. Art New, Jr., of Edmonton displays a 15-pounder. He and partner Ron Close, bagged their five-bird limits within half an hour this week. Four of their Canada geese were within a pound of this handsome bird."

Large Canada geese, according to R. Webb, "breed throughout Alberta with greatest densities along the rivers and near irrigation areas of the southern prairies. No adequate census has been taken but the total would probably approach 10,000 birds in fall. Densities have been increasing in recent years."

Dzubin (*op. cit.*) has also furnished detail on the distribution of Canadas in Alberta:

> In the few spring census flights made into central Alberta, into the Hanna district, a few pairs of Canadas are found on almost all lakes there. I would suspect the breeding area of large Canadas to be east of a line from Edmonton, Red Deer, Calgary and Lethbridge. On April 16, 1961, a census flight in the Hanna district indicated 1 to 3 pairs breeding on each of Pearl, Chain, Sullivan, Shooting, Buffalo, Driedmeat, Bittern and Wavey Lake. Some 5 pairs were noted on Dowling Lake, northwest of Hanna and

15–20 pairs on islands and points of Farrell Lake, northwest of Hanna. On April 17th in the district south of Hanna we counted 2 pairs on Coleman Lake, 2 pairs on Sunnynook Dam, 5 pairs on Hardhills reservoir, 2 pairs on Little Fish Lake, 3 pairs on Sieu Lake, 3 pairs on Mattoyeku Lake, 2 pairs on large sloughs north of Duchess, Alberta, 1 pair on Berry Lake, 1 pair on Plover Lake and 1 pair northeast of Stanmore (in each case a lone bird that refused to flush was considered an "indicated pair"). A flight along the Red Deer River with Tom Sterling from a point south of Dorothy to the junction with the South Saskatchewan River on the Saskatchewan border gave us an indicated population of nearly 100 pairs. They breed on the S. Saskatchewan River with the major breeding grounds between the two systems — S. Saskatchewan and Red Deer — mainly in the Brooks-Bassano irrigation districts. . . . A partial lake census into central Alberta on May 1, 1961, indicated 2 pairs of Canadas on Sounding Lake, 2 on Kirkpatrick, 1 pair on Dead Horse Lake, 6–8 pairs on Dowling Lake, 2 pairs on Chain Lake, 12–15 pairs on Farrell Lake, 2–3 pairs [in the area] surrounding Sullivan Lake, 1 pair on Lanes and Cutbank Lake, 5–7 pairs on Shooting Lake, 2 pairs on Manon Lake, and 5–6 pairs on Buffalo Lake.

United States

IDAHO — A population of Canada geese breeding in Canyon County, Idaho, appears to be an important exception to the generalization that the breeding range of *maxima* lies east of the Continental Divide. Skins of geese from this population in the Carnegie Museum show that it is comprised of very large birds (Table 12) which possess the culmen and tarsal characters of *maxima*. In view of this finding, a collection of heads of Canada geese from Idaho, which had been obtained for me earlier for a speciation study, were re-examined. In respect to both size and shape of the culmen, the heads of the Canada geese from Canyon County in the southwest portion of Idaho were found to be much more like *maxima* than the collection from Blackfoot Reservoir in Caribou County which were identified as *moffitti* (Table 13, Figs. 14–16).

The recent history of the Canada geese in Canyon County, the Homedale flock, is of interest because of its apparent subspecific identity. According to Elwood G. Bizeau (*letters* of June 6 and July 11, 1963) of the Idaho Fish and Game Department, the Homedale flock is scattered along the lower Snake River from Glenns Ferry to Brownlee Dam below Weiser, including tributary streams. Three thousand to four thousand goslings are produced annually in the area, the total flock being about twice this number. In autumn, this population gradually concentrates at Lake Lowell on the Deer Flat National Wildlife Refuge; by the opening of the hunting season, 6,000 to 8,000 geese are present. Later in the hunting season there is an increase of about 1,000 birds which Bizeau believes may represent geese from more outlying areas or migrants from Canada, as suggested by a limited number of recoveries of geese banded in Canada.

It is significant that the Homedale flock is resident in the area and does not migrate southward as do the flocks breeding in southeastern Idaho. All first-year recoveries of banded goslings from the Homedale flock are taken within 75 miles of the banding site. This is also true of recoveries of geese in subsequent years of life, although a few yearlings are shot in Canada and eastern Washington. This northward dispersal of yearling giant Canada geese is discussed in detail in Chapter 4, "Migration."

The racial identity of the Homedale flock is by no means a "closed book"; additional specimens should be collected and the flock, as a whole, carefully scrutinized in light of data on the characteristics of the race presented here. The Carnegie Museum skins indicate that the Homedale flock is a dark-colored population.

The origin of what now appears to be a pocket population of *maxima* must remain conjectural. It may have originated from stocks from Alberta that follow portions of the Snake River when migrating to northern California. Inquiries of Yocum (1962:8) determined that there were no significant numbers of Canada geese along the Snake River in Idaho as late as 1910. With the advent of irrigation and agriculture along the Snake River, the number of Canada geese breeding in this sector of Idaho rapidly increased.

MONTANA — The eastern two-thirds of Montana lies within the prairie biome; therefore, the Canada goose populations breeding in this area could be expected to be *maxima*. Information and photographs (Fig. 31) received from Dale Witt have supported this conclusion. The largest goose weighed by Witt in the vicinity of the Bowdoin National Wildlife Refuge during the hunting season was 14 pounds and he reported that many geese were in the 11 to 12-pound range. Regarding the coloration of Canada geese in Montana, Witt writes (*letter*, April 19, 1963): "I have seen big geese similar to the description for *maxima* throughout the different flocks in Montana during the banding operations. The coloration of these large geese seems to be most outstanding, particularly the white on the forehead and white collars, but the number of geese with these characteristics is very small comparatively."

According to Dale Witt (*letters*, April 19, 1963, and December 6, 1963), there are three principal population units of Canada geese in

Montana east of the Continental Divide: The Hi-Line Unit, which includes the flock on the Bowdoin National Wildlife Refuge, occupies an area extending northward from the Fort Peck Dam and westward along the Missouri River to near Chinook and is contiguous with populations breeding in southwestern Saskatchewan in the Cypress Hills–Swift Current area; the East Slope Unit, which nests in the vicinity of lakes and reservoirs in the Cutbank region and is contiguous with the populations in the Brooks-Bassano-Bow River district of Alberta; and the Helena Unit, which is largely concentrated in the vicinity of Canyon Ferry Reservoir and Lake Helena. There is also a small population at Medicine Lake in extreme northeastern Montana. The scattered population along the lower Yellowstone River is also probably *maxima*. Breeding populations of Canada geese in Montana west of the Continental Divide would, in all likelihood, be *moffitti*.

Since 1960, the Montana Department of Fish and Game has made counts of Canada geese on the major lakes and reservoirs in the breeding area of each of the three units. These counts are taken annually to show population trends; the actual total number of geese in each population unit is somewhat higher than the inventory counts which do not constitute a complete census. Nevertheless, the data obtained are of value as they give the minimal size of the Montana population (Table 15).

15. Minimum numbers of giant Canada geese in Montana east of the Continental Divide, summer 1962

Population unit	Yearlings and adults	Immatures	Total
Hi-Line	1,010	2,079	3,089
East Slope	478	317	795
Helena	276	348	624
Total	1,764	2,744	4,508

COLORADO — According to Jack R. Grieb (*letter*, April 16, 1963), little factual information exists regarding nesting Canada geese in Colorado in early years. Cooke (1897:58) wrote: "On the plains of eastern Colorado they are known only as migrants and winter residents. In the mountains they breed along the higher secluded lakes at about 10,000 feet, especially in North Park, where Dr. Coues found them breeding in large numbers." In an appendix to his initial report, Cooke (1898) reported Canada geese nesting 5 miles west of Niwot, Boulder County, at 5,500 feet. In a second appendix, Cooke (1900) stated that they bred in Middle Park.

Grieb has talked to an "old-timer" in North Park who stated that they used to club geese to death on the North Platte River, presumably when the geese were in molt. He also reported that there were records of nesting Canada geese from San Luis Valley. There are some data, however, that initially indicated to me that the subspecies of Canada goose breeding in eastern Colorado in earlier times was *maxima:* the largest eggs of which I have knowledge were collected in Colorado (Sclater, 1912 and Table 19), and the weights of three birds shot in Colorado in 1889 and reported to Mershon (Table 1) would identify them as *maxima.* It is possible, however, that these geese could have been migrants.

Purely on ecological grounds and the absence of geographical barriers, I had considered, *a priori,* that any Canada geese that may have nested in the portion of Colorado drained by tributaries of the Missouri River had to be the same subspecies as that which nested on the plains of the Dakotas, Nebraska, and Kansas. New evidence for this opinion was found in a series of photographs of nesting Canada geese near Fort Collins which were published in the Sunday magazine section of the *Denver Post* in May, 1963. The long neck, size and slope of bill, and white collar present on many of the birds in the photographs left little question in my mind that these geese were indeed *maxima.* Conclusive determination of the identity of the Canada geese nesting in the Denver and Fort Collins areas was made possible in October, 1963, when I visited the Denver area, under the sponsorship of the Denver Museum of Natural History, and was able to observe the local flocks (Fig. 32). On this occasion, five large adult males were collected for my study and for the museum's skin collection. These males proved to be magnificent examples of the race *maxima.* The heaviest goose was selected for size, but the other four are more nearly representative of the flock. The measurements and weights of these specimens are as follows: wing, 536.2 mm (520–551); tail, 161.0 mm (156–168); culmen, 60.4 mm (55–65); tarsus, 99.0 mm (94–102); middle toe and claw, 100.8 mm (98–108); and weight, 13 lb. (11 lb. 4 oz to 14 lb. 14 oz).

While in the Denver area, Jack Putnam, of the Denver Museum of Natural History, and I were able to count about 600 giant Canadas, but Jack Grieb has assured me that the local flock, which uses both the city park and the lakes and reservoirs of the surrounding countryside, totals at least 1,000 birds. Grieb has commented further on the status, management, and origin of the giant Canada goose in the Denver and Fort Collins areas (*letter,* November 12, 1963):

In addition, we have about 550 birds in the Fort Collins area located on College and Terry Lakes. College Lake is four miles west of Fort Collins, and Terry about two miles north. These birds concentrate on these areas in the winter, but spread to many small waters during the breeding season. We operate an aerator at College Lake to keep the water open during the winter, but the birds are able to do the job themselves on Terry probably because quite a few ducks are also on this lake. We feed at both areas to hold the birds, and reduce losses during the hunting season. We have almost reached the stage where some harvest can be made on this flock, but we intend to approach this carefully.

We have not taken all the young geese produced by the Denver flock [for transplant], but we have taken all geese produced by the small segment at Denver City Park. In addition, we take 50 eggs from nests at Bowles Lake and hatch these with hens or incubators for release. This year we released 86 birds at Valmont Reservoir at Boulder, Colorado.

The Bowles Lake geese, and I assume most of the other breeding Denver geese originated from about a dozen birds kept at Bowles Lake by Dan and Virgie Gallagher. They still live on the south side of the lake and feed these birds constantly. The way Mrs. Gallagher told it, they were at a friend's house and he was going to butcher his captive, decoy goose flock when this type of hunting was outlawed. She bought them from him and released them on Bowles Lake. I would suspect that some of the Bowdoin birds stopped during their migration and became part of this flock, and I'm sure that some of these Bowdoin birds winter in Denver since the number of geese goes up considerably during the winter.

This breeding flock does not migrate away from the Denver or Fort Collins area although we may lose a few birds to the migratory flock which winter or pass through this area.

NORTH DAKOTA — As pointed out earlier, the principal range of *maxima* in North Dakota, as well as adjacent South Dakota and Minnesota, coincided with glacial Lake Agassiz, and most of the present-day nesting populations are confined to the region formerly occupied by these glacier meltwaters. The area around Dawson, Kidder County, was famous in early days for the duck and goose hunting it afforded; from the standpoint of hunting, it might be considered the type locality of *maxima* (Fig. 33). The early fame this area achieved as a hunting center was probably due in part to its being accessible by rail.

In the early 1890's, the Canada goose was regarded as a common breeding bird in the general region of Caddo, North Dakota (Judd, 1917:9). By 1902, Bent (p. 173) had already found a decrease in numbers:

> There are still quite a number of Canada geese breeding within the limits of North Dakota, but they are apparently not as abundant as formerly and will undoubtedly be driven further west and north as the country becomes more richly settled. We found several of their nests [presumably in Nelson County], but for some unaccountable reason we did not see a single goose.

Later Bent (1925:206) wrote: "When I visited North Dakota in 1901 there were still quite a number of Canada geese breeding there."

Wood (1923:22) has documented the early records of Canada geese in North Dakota. In 1923, he considered it to be a rare breeder in the state, but reported that it still nested at Sweetwater Lake, Ramsey County. In 1920, he observed a family flock on Devils Lake.

Letters in the Mershon and Holland files chronicle the elimination of the Canada goose as a breeding bird in North Dakota.

L. C. Pettibone of Dawson to W. B. Mershon, June 23, 1923: "I am told there are a few of them nesting 25 miles south east of here."

L. C. Pettibone to W. B. Mershon, November 12, 1923: "There was a brood of them hatched a few miles north of here [Dawson, Kidder County]."

E. T. Judd to W. B. Mershon, December 24, 1924: "The Big Canada Goose I am afraid is gone, altho I have one pair at our Fish Hatchery in Turtle Mts. and I sent one pair to Washington Zoological park. 2 pairs nested near Lake Williams in 1923, but did not learn of any nesting in 1924."

By 1924, the Canada goose as a breeding bird must have disappeared from most of North Dakota as Lincoln (1925), in his study of the summer waterfowl, makes no reference to this species. Yet a few evidently persisted: "there are still a few of the big geese that nest here but I have been unable to get any." (L. C. Pettibone to W. B. Mershon, January 15, 1926, in response to the latter's effort to obtain specimens.)

The disappearance of these geese in the wild was evidently almost complete by the 1930's.

R. P. Holland to John C. Phillips, November 7, 1934: "I was talking to an old gunner out in Saskatchewan this fall and without a word from me he said, 'We never see any of the big geese any more. I guess they are all gone. I mean the big grey fellows who weighed from 16–18 pounds.' This man was born and raised, I think he said, in one of the Dakotas."

North Dakota strains of *maxima* were saved from extinction by game breeders, many of whose flocks had been formerly used for decoys. Descendants of these flocks are now being used to restock federal refuges in

North Dakota. E. I. Smith of the North Dakota School for the Deaf at Devils Lake has several pairs of exceptional size; 23 offspring of these geese were at the Sully's Hill National Game Preserve at Devils Lake in the fall of 1962. Carl E. Strutz, a waterfowl hobbyist and game breeder at Jamestown, North Dakota, has several dozen of these large geese, two of which when 2 years old in 1963 weighed 20 pounds each. Many of the geese currently on federal refuges in the Dakotas are descendants of an old 18-pound male (Figs. 9 and 34) and its mate owned by Strutz. In 1962, nine offspring from this pair were being held at the Lower Souris National Wildlife Refuge for restocking purposes (Fig. 10). This same year there were 68 giant Canada geese at the Snake Creek National Wildlife Refuge (Fig. 37), an area which was developed following the Garrison Dam impoundment of the Missouri River. This flock is derived from a number of sources: a game breeder in south-central North Dakota; in Michigan, the Seney National Wildlife Refuge, a private game breeder, and the State Game Farm.

At Lostwood National Wildlife Refuge (Fig. 35), there were 35 two-year-old giant Canada geese in the fall of 1962, offspring of pairs owned by Carl E. Strutz.

At the Upper Souris National Wildlife Refuge there were approximately 160 giant Canadas in the fall of 1962. In part, this flock is descendent from stock obtained from the Waubay National Wildlife Refuge, South Dakota.

SOUTH DAKOTA — I have found little information on the early history of the Canada goose in this state. By 1906, Cooke had reported "that it now breeds rarely . . . in southern South Dakota." It is reasonable to suspect that prior to settlement the range of *maxima* in South Dakota was confined mainly to the area east of the Missouri River, and that the race was most numerous in the lake country in the eastern tiers of counties and the area of the state formerly occupied by glacial Lake Agassiz.

The largest wild population of *maxima* in the Dakotas in the fall of 1962 was 380 in Day County, concentrated at that time at the Waubay National Wildlife Refuge (Fig. 33). Day County offers tremendous potential for the restoration of these geese, and all public agencies concerned should make a concentrated effort to restore the giant Canada goose to this area in as large a number as the region will permit. As a tourist attraction and, subsequently, as a game bird, the economic gain alone should justify the effort.

In 1961, the Sand Lake National Wildlife Refuge held 326 large Canadas. In 1939, western Canada geese (*moffitti*) had been released on this refuge to initiate a local breeding population, but the weights of Canada geese shot in the surrounding area in more recent years and the habits of the geese now nesting on the refuge strongly suggest that the original stock has been supplanted by locally nesting *maxima*. A single crippled immature examined by me in October, 1962, was *maxima*. In early November, 1947, an 18¾-pound goose was shot near the north end of the refuge (Fig. 8), one of a flock of four flying low in a snowstorm. In most years the flock at the Sand Lake refuge produces about 75 young, but in 1962, production was considerably less as many nests were lost to flooding. An extensive program recently inaugurated to provide safe nesting platforms (Fig. 70) for this flock should substantially increase its productivity.

NEBRASKA — The giant Canada goose was exterminated from most of Nebraska before the turn of the century; however, with the possible exception of the Sandhills, the population in the state may never have been large. In 1896, Bruner, citing D. H. Talbot, reported that the Canada goose occurred at "numerous localities in the central part of the state." A few years later Bruner, Wolcott, and Swenk (1904:30) wrote: "Formerly bred about the lakes in the sandhill region and on the islands of the Platte and Missouri rivers, and a few are found breeding there still." In 1906, Cooke (p. 72) wrote that the Canada goose "now . . . breeds rarely in Nebraska." Haecker, Moser, and Swenk (1945) have also commented in the same vein: "Formerly bred in the sandhills area and along the larger rivers of the State."

Years later, in a letter to R. P. Holland (November 13, 1934), Swenk emphasized the lack of knowledge concerning early nestings of Canada geese in the state: "We know that in the early days in Nebraska, Canada geese bred along the streams of the state; however, no specimens of these locally breeding geese have been preserved, so far as we know. Probably these local birds were extirpated very early, and only the early-day hunters have had any personal contact with them."

In recent years, Canada geese have been re-established in the Sandhills at the Valentine National Wildlife Refuge. The original stock was purchased from game breeders in Nebraska. In 1962, a flock of Canada geese at the Crescent Lake Wildlife Refuge produced 100 goslings. The breeding geese at this refuge originated from former decoy flocks that were present in the northwest quarter of the state. The city park in North Platte also contains a flock of giant Canada geese, descendent from old decoy stocks, which has been producing 300 to 400 young per year (Harvey W. Miller, *personal communication*).

KANSAS — Little seems to be known about the early status of Canada geese in Kansas. Huntington (1910) and Delacour (1954) both include Kansas in the former breeding range of the species. It evidently disappeared from the state at an early date as, in 1906, Cooke wrote, "formerly bred in Kansas." Probably most nesting in the state was adjacent to rivers, but Leo Kirsch has suggested (*personal communication*) that the Cheyenne Bottoms in Barton County and the McPherson Bottoms west of McPherson may have been important nesting grounds prior to settlement.

MINNESOTA — As a result of the studies by Roberts (1932) on the birds in Minnesota, the distribution of Canada geese in this state in earlier times is well documented. Although he stated (p. 208) that it formerly nested throughout the state, it seems doubtful to me that the Canada goose ever nested in any numbers in the Superior Uplands of the northeast quarter of Minnesota. As late as 1906, Cooke (p. 72) reported that "throughout much of Minnesota the species is a regular and not uncommon summer resident."

Roberts' (1932) records of nesting Canada geese in Minnesota have, in effect, documented the gradual reduction of the species in the state as a breeding bird: Marshall County (Mud Lake), 1891 and 1901; Cass County (Leech Lake), 1902; Sherburne County (Elk River), 1884, 1888; Jackson County (Heron Lake), 1894; and Nicollet County (Swan Lake), 1923, 1925, 1927, 1928, 1929, and 1930. The late Eli Boudrye of Granada, Minnesota, reported that he found nesting geese in Martin County about 1890; prior to 1900, settlers near Swan Lake, Nicollet County, robbed nests for eggs and captured young which were eaten in the fall (Forrest B. Lee, *letter*, November 19, 1962).

An interesting early account of Canada geese in Minnesota is contained in a letter of Morton Barrows to W. B. Mershon, August 2, 1922:

> I came to Minnesota in 1883, but did not begin shooting geese until several years after that. Most of my goose shooting was done in the Counties of Swift and Yellow Medicine. From 1885 to 1890 geese were fairly plentiful in both of these counties, although I am now quite inclined to the opinion that they were very largely local birds. In those days I thought very little of species or exact weights, and although I killed a great many very large geese, which I am sure weighed over twelve pounds, I can recollect weighing only one. That was killed about 1888 as nearly as I can place the date. My father and I were shooting just south of Murdock in Swift County. We killed one goose that I put on the scales after I reached home and which weighed something over sixteen pounds. I cannot recollect that it differed in markings or color from the other geese. . . . Some years ago at Heron Lake, Minnesota, . . . I saw a stuffed goose which people there

> told me weighed twenty-four pounds, and certainly the stuffed specimen looked to be fully that large. It was simply enormous. . . . but goose shooting has as you know long been a thing of the past, except in the Thief River region of Northern Minnesota. This, I think, is due more to drainage and the settling up of the country than to extermination.

Today there are three major resident populations of *maxima* in Minnesota: at the Agassiz National Wildlife Refuge, the free-flying flock was 550 (in 1962; 800, 1964) and gosling production, during the 1955–61 period, averaged 115 (Nelson, 1963) (Fig. 34); a flock of 170 at the recently established Lac Qui Parle State Refuge (Fig. 34); and a captive, free-flying flock of several hundred at the Round Lake Waterfowl Station. The latter flock is derived from game breeders' stocks in Minnesota, Iowa, and the Dakotas. There are also two smaller flocks on estates near St. Paul (the Hill and Bell properties). The Hill flock originated from birds obtained from a game breeder in Rock County, Wisconsin, and from the late Hans J. Jaeger, Owatonna, Minnesota. It is now free-flying and independent.

The Agassiz flock has a particularly interesting lineage and history. The initial release of geese (*moffitti*) on the Agassiz refuge (Fig. 36) came from the Bear River refuge, but all later releases were of geese obtained from the Seney National Wildlife Refuge in the northern peninsula of Michigan. According to Manly Miner (*personal communication*, 1962) of Kingsville, Ontario, the Seney geese were derived from a flock formerly owned by Henry Wallace of Milford, Michigan. He, in turn, on the advice of the late Jack Miner, is said to have obtained his original breeding stock (three pairs) from the late Hans J. Jaeger, a game breeder in Owatonna, Minnesota. However, in a letter to W. B. Mershon (March 22, 1924), Jack Miner reveals that other blood lines were incorporated into this flock. "But a Mr. Wallace of Detroit, Mich. took me out to his place to see his wild geese that were wing tipped and brought from North Dakota and these geese are fully 20 lb., of a larger frame than those that migrate through here [Kingsville, Ont.]."

Thus, the Agassiz refuge *maxima* may have a dual lineage, but the present flock is unquestionably descendent from the large-sized stocks of *maxima* that ranged through southern Minnesota and North Dakota. (*See* the measurements of specimens from the Agassiz refuge area [Marshall County] and from the Seney refuge in Table 7.)

All but 5 of 135 pinioned geese present at the Lac Qui Parle State Refuge in the fall of 1962 were transplants from the Carlos Avery State Refuge (Fig. 34). These five were purchased from Emmet E. Kern, a farmer and game breeder near Granada, Martin County, Minnesota; these are particularly large geese, especially an adult male (Fig. 9). Mr. Kern received some of his first geese from Lester Case of St. Cloud, Minnesota, and from Mr. Tibisar of Rollingstone, Minnesota, but reports his largest geese were obtained from Bert Welcome, Sherburn, Minnesota (*letter*, December 15, 1962).

At Thief Lake, a flock of pinioned giant Canada geese is being held for restoration of this race in that area. These geese originated from Carlos Avery stock and from goslings salvaged from refuges in Minnesota and Wisconsin. This flock may contain a few specimens of *interior*.

IOWA — Anderson (1907) has summarized the early history of the Canada goose in Iowa: "Before the general settlement of the state the Canada Goose nested quite commonly in various parts of the state and a few pair still linger throughout the summer in localities which are not too thickly settled [p. 185]." He has cited the records of Lewis and Clarke and other early explorers, pointing to the fact that the Canada goose formerly must have nested commonly along sections of the Missouri River bordering Iowa. It once nested in the Spirit Lake area and commonly in Winnebago and Hancock counties. Anderson listed as a typical specimen from Iowa a male shot by D. H. Talbot near Sioux City on November 17, 1886. This goose had a wing spread of 72 inches and a wing length of 21.5 inches (546 mm); a tail of 7.25 (184 mm); bill, 2.44 inches (62 mm); and a weight of 13.5 pounds (6,123 gm). This specimen was unquestionably *maxima* and, judging from length of wing and tail, was an adult male.

Cooke's (1906:72) assessment of the former range of the Canada goose is pertinent here: "A hundred years ago the species bred commonly in all the northern third of Mississippi Valley and not uncommonly to the latitude of St. Louis. Now the number of pairs breeding south of the latitude of central Iowa is very small."

MISSOURI — According to Wiedman (1907:46), "When the first white men flocked into the state, they found the geese nesting all along the Mississippi and Missouri Rivers. . . . During the last decade of the past century the Peninsula of Missouri [currently referred to as the bootheel — Mississippi and Scott counties] still harbored a small number of breeding pairs, usually nesting on cypress stumps in the overflow 6 or 8 feet above the water."

In a recent letter (March 13, 1964), M. G. Vaiden of Rosedale, Mississippi, sent me an account of Canada goose nesting in the bootheel of Missouri near the turn of the century.

> I have very recently finished a conversation with Mr. Frank England, a gentleman 77 years old, and who was born in Mississippi County, Missouri, within two miles of the Mississippi River.
>
> Mr. England tells me that when he was 9 or 10 years old he went with his father and older brothers to rob Canada Geese nests in Jim's Bayou where there was an enormous growth of cypress trees. These trees had the tops blown out from the earthquake that created Reelfoot Lake and/or from the winds that followed the "quake." Very large Canada Geese would nest in the tops of these cypress trees and that he had to climb up to many nests and had stolen the fresh eggs for they were cooked and used at home. Some members of the family believed a goose egg fried was better than a hen egg. That too, they could not catch, from a skiff, any of the little birds from the water for they were great divers but that when the first feather change came [molt] they could not dive and that they would then catch these geese and take them home to be "tamed" for future eating. He remembered that at one time in his youth he had about twenty of these yearlings in the pens with the chickens and ducks. Mr. England also stated that the cypress brake forming Jim's Bayou was a very large one and that all along the bayou settlers in this section of Missouri's bootheel used the same methods. Too, that when he went back at the age of about twenty-five years for a visit there were no big geese. Mr. England states, "These were real geese and greater in size than the usual run of geese."
>
> Mr. England pinpoints the area as lying between East Prairie and the Mississippi River.

The recent study by McKinley (1961) on the history of the Canada goose in Missouri confirmed my belief that early records of such a conspicuous bird as the Canada goose could be greatly amplified if accounts of the early explorers, pioneers, and county histories of the north-central states were carefully searched. For example (p. 5):

> That geese nested near New Madrid before the great earthquake is proved by the report of Christian Schultz (1810, 2:18) who visited the town in 1807. People, he was told, "frequently catch the young goslings before they can use their wings, and rear them with a tame brood."

Although McKinley has concluded that "it seems possible that the Canada goose originally nested, at least at times, over much of our state," his references (with one exception, Miller County in south-central Missouri) and my own concept of the range of this subspecies are not in agreement with this statement. Records he cites are all from along the Missouri or Mississippi rivers, the alluvial soils of which are comparable in fertility with the northern prairies. Much of Missouri is too rough in topography and too heavily wooded to be attractive to Canada geese.

The one area of Missouri, other than the linear range of river bottoms, that might have supported a thinly scattered population of Canada geese is the north-central prairie region; however, records to substantiate this hypothesis are not known to me.

A nesting population of *maxima*, which in 1963 totaled 320, has been established on the Trimble Wildlife Area near Trimble, Clinton County (Vaught, 1960).[4] This population, which now also breeds in the surrounding countryside, originated from an old private flock in Boone County. Washtubs placed in trees for nest sites have increased nesting success (Brakhage, 1962) (Fig. 72 and Table 22).

ARKANSAS — Bent (1925) lists *Branta canadensis canadensis* as formerly nesting at Walker Lake in northeast Arkansas. This record for Arkansas is presumably based on Howell (1911:22): "a few pair remain to breed in the most secluded parts of the Sunken Lands [of Arkansas]. At Walker Lake on May 4, 1910, I saw a pair and was told that several pair breed there each season."

WISCONSIN — Kumlien and Holister (1903:28), reporting on the early status of the Canada goose in Wisconsin, wrote:

> Fifty years ago a common breeder in almost any swamp or large marsh, or on the "prairie sloughs" (now a feature of the past). At the present time only scattered pairs nest as far south as the southern third of the state. The last nesting record we have for southern Wisconsin was in Jefferson County — from the years 1891–99, inclusive, when a goose deposited her eggs on the edge of a tamarack swamp, on the same mound of rubbish each year. The first set was taken several times, when she moved to another mound farther into the swamp and here hatched her eggs. No mate was ever noticed to have visited her.

In his studies of accounts of wildlife published in early newspapers and other sources in Wisconsin, Dr. A. W. Schorger found a number of historical records of unusual interest. One of these was from the *Whitewater* [Wisconsin] *Register*, May 31, 1888: "A party who spent a day there the last part of the week saw a wild goose which is evidently summering with her mate at the lake, and she was so tame she did not take flight until they had approached within a few rods."

Main (1943:214) provided an interesting record in her biography of Thure Kumlien in which she cited his letter to Thomas M. Brewer, written from Lake Koshkonong on January 19, 1852: "When I spoke of Goose eggs I did mean A. Canad. [*Anser canadensis* = *Branta canadensis*] and I have lately found out another of their breeding places where they be more numerous."

In a biography of Philo Romayne Hoy, Schorger (1944:58) wrote:

"Hoy's journals and unpublished notes have not been seen. Judging from the few extracts that have appeared (Racine *Journal-News,* May 2 and 24, 1922), they contain valuable information. . . . The Canada goose at that time nested abundantly. In 1850 their eggs were gathered by the bushel in a marsh north of Racine."

In northwestern Wisconsin, a population of giant Canada geese nested on muskrat houses in shallow Munson Lake, Burnett County, as late as 1911 (Hollis Barrett to N. R. Stone, *personal communication*). Mr. Barrett recalled seeing from 20 to 25 pairs of geese on Munson Lake each summer between 1905 and 1911. When wild hay was cut in the marsh, the geese were disturbed and their heads could be seen projecting above the vegetation all around the lake. As late as May, 1942 or 1943, six geese were observed in the pot-hole area of Crex, north of Riesmeyer Lake. These geese apparently sought to nest, but were driven off by poachers.

Mr. Barrett also reported to N. R. Stone that he was certain a pair of geese nested on Forman Lake in the Crex area in either 1938 or 1939. On the opening day of fall hunting, hunters shot five or six geese on this lake, presumably the adults and their young of the year.

Sometime after 1911, the Crex marsh was ditched and drained and the local breeding flock of wild Canada geese was eliminated. This marsh area is now being restored for wildlife by the Wisconsin Conservation Department; Crex Meadows Wildlife Area as it is now known, and which includes Munson Lake, will comprise 24,500 acres when acquisition and restoration are completed. Giant Canada geese purchased in the winter of 1949–50 from various game farms have been re-established in the area. Some of the pairs purchased came from the Inman Farm flock on Rock Prairie, which I have observed and identified as *maxima.* According to Richard A. Hunt, the present flock at the Crex Meadows Wildlife Area is also descendent, in part, from birds originally obtained from a game farm in Nebraska and from pairs that were purchased from the Bright Land Farm, Barrington, Illinois.

A second breeding flock of giant Canadas, now totaling approximately 200 birds, has been built up over a period of years at the Louis Barkhausen Game Sanctuary near Green Bay; a third flock of about 100 birds is resident at the Bay Beach Wildlife Area, Green Bay. I observed this latter flock in early July, 1962; its members are similar to the Manitoba populations, some showing to an extreme degree the plumage characteristics of the race.

Recent breeding records and occurrences of Canada geese in Wisconsin are shown in Map iii.

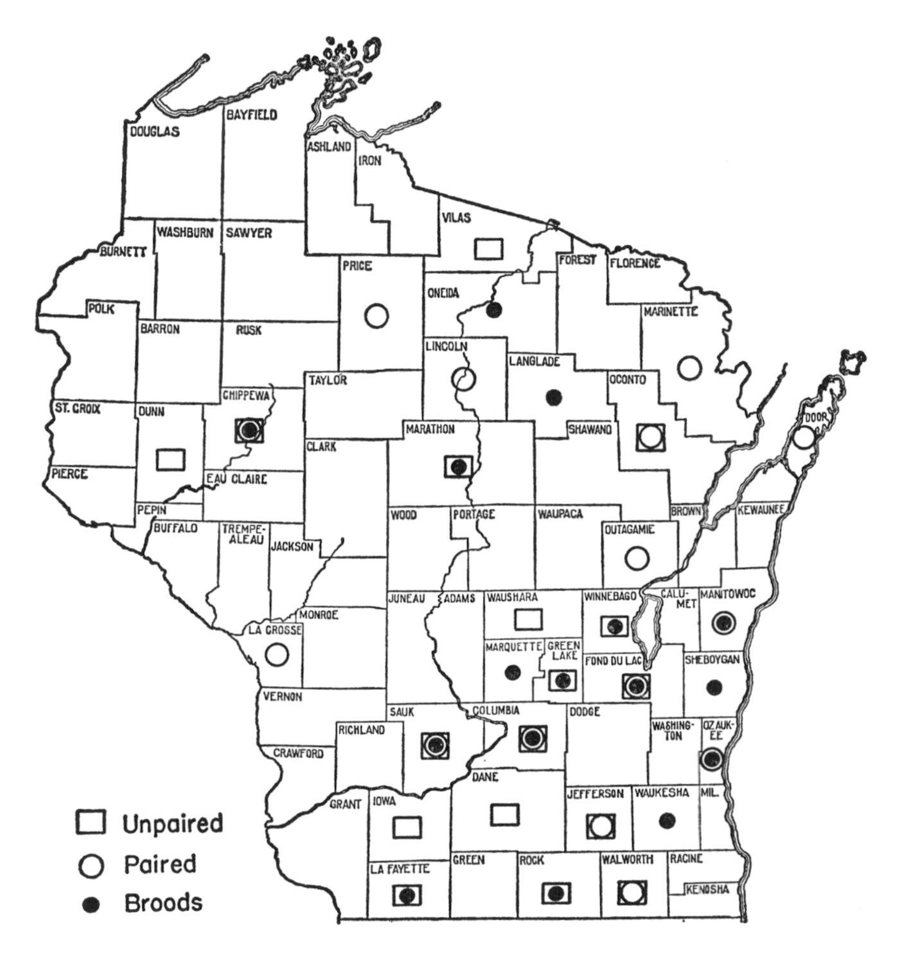

iii. *Records by counties of Canada geese observed in Wisconsin between June 1 and September 1, 1947–1963. Excluded are records from areas in the vicinity of breeding flocks at Horicon, Neceedah, Crex Meadows, Goose Island, Powell Marsh, and Green Bay. Brood records are assumed to be of* Branta canadensis maxima. *(Data courtesy of Richard A. Hunt.)*

ILLINOIS — Woodruff (1907:50) reported that the Calumet marshes once formed the breeding site of the Canada goose in the Chicago area. He cites Nelson's (1876) statement that the Canada goose "formerly bred commonly in the marshes throughout the state, and still breeds sparingly in the more secluded situations." This appraisal of the range in Illinois should probably be amended to exclude the hilly and heavily wooded counties in the southern portions of the state, except those areas bordering the major rivers — the Wabash, Ohio, Kaskaskia, Big Muddy, and Mississippi.

One of the earliest collected specimens of the giant Canada goose I have studied was collected for Robert Kennicott in 1864 at Chicago (Table 7). It is now in the Philadelphia Academy of Natural Sciences.

An early nesting record of Canada geese in Illinois, called to my attention by Dr. A. W. Schorger, is remarkable for the amount of specific information it contains. In an account of muskrat trapping in 1845 at Mud Lake, 12 miles from Chicago, Henry Thacker (*in* Newhouse, 1874:157–58) tells of shooting a Canada goose:

> One day as I was pushing my little boat along through the tall reeds, I saw at a distance something unusual on the top of a muskrat house. As it was lying flat, almost hidden from view, I at first sight took it to be an otter as we had killed one some time previous near the same place. As usual at the sight of game, my rifle was quick as thought brought to bear, and away sped the bullet, and over tumbled a large wild goose, making a great splashing as she fell into the water. On examination I found she had a nest of seven eggs, all fresh. The goose weighed fourteen pounds and a half. The same day I found another nest with several eggs, and took them to a farmer who was anxious to get them to hatch "at the halves."

During the course of this study I realized for the first time that the various captive and semicaptive flocks of Canada geese in Illinois are *maxima*. Until recent years, the largest and most successful nesting flock in Illinois has been held on the Bright Land Farm near Barrington, Illinois. In the mid-1940's, this flock numbered about 250 birds, a fourth of which were free-flying (Kossack, 1950). The original breeding stock for this flock was obtained from a game breeder in Wisconsin. Birds from the Bright Land Farm formed the original nucleus of a flock of several hundred birds now held on the Robert Bartlett estate a few miles north of Barrington.

In the 1950's, Glenn Palmer, former Director, Illinois Department of Conservation, purchased captive Canada geese for stocking the state refuge and game farm on the Des Plaines River near Joliet from the following game bird breeders: the Bartlett estate near Barrington; Shirley Brigham, Hebron, Illinois; C. F. Kerfoot, Tennessee, Illinois; Tanner Glass, Galesburg, Illinois; and a Mr. Logsdon, Cooperstown, Illinois. I examined 84 immatures produced on the Des Plaines area in 1962. These proved to be some of the largest young I have studied (Tables 9 and 11).

MICHIGAN — According to Barrows (1912:118), "Formerly they doubtless nested more or less commonly all over the state, and it is not impossible that single pairs may do so still in favorable places." Barrows also stated that others thought that "they bred sparingly in the neighborhood of Monosco [Monoscong] Bay and Hay Lake, St. Mary's River."

The Seney National Wildlife Refuge now has a free-flying population of 1,100 Canada geese, the great bulk of which, in light of circumstances detailed under "Minnesota," are originally from *maxima* stock. This was confirmed by personal study of live specimens made in July 1963. The re-establishment of giant Canada geese on the Seney refuge emphasizes the importance of reintroducing to an area the subspecies which was adapted to it. However, this concept was not realized as the basis for the transplant program at the time, as crippled wild birds (*interior*) from Horseshoe Lake and other wintering areas were shipped to the Seney Refuge in the 1940's.

The success reported by Pirnie (1938) in establishing breeding geese in southeastern Michigan with geese raised at the Kellogg Refuge is indicative that this stock also consisted of *maxima*. (The nesting Canada goose shown in a photograph in Pirnie's 1938 report is *maxima*.)

One of the most notable successes in restoring giant Canada geese to the Midwest in recent years has been achieved at the Shiawassee National Wildlife Refuge in Michigan. Between 1959 and 1962, 360 Canada geese were released on the Shiawassee refuge. In 1962, 60 nesting pairs produced approximately 225 goslings (Nelson, 1963). The stock released was obtained from the Michigan State Game Farm at Mason which, judging from geese of this origin seen elsewhere, are unquestionably *maxima* (*see* photograph of pair in Martin, 1963). Again, as in the vicinity of the Kellogg and Seney refuges, success was easily attained because the stock used was the endemic strain adapted to the region. For the first 3 or 4 years the breeding flock at Shiawassee did not migrate, but in the winter of 1963 they were absent from the area for about 4 weeks; it is believed that they wintered in southeastern Michigan (Harvey K. Nelson, *letter* of August 15, 1963). The sedentary behavior of the Shiawassee flock is indicative of its racial identity. Between Pontiac and Flint, Michigan, there is also a flock of several hundred giant Canadas. The significance of the nonmigratory behavior of *maxima* will be pointed out in Chapter 15, "Discussion."

INDIANA — In 1897, Butler (p. 637) reported that Canada geese "still breed in some numbers in the Kankakee region and less frequently in other favorable localities, notably DeKalb County . . ., Steuben County . . ., at Twin Lakes of the Wood . . ., Laporte County." The Kankakee River marshes were possibly one of the most important breeding areas of the giant Canada goose in the Midwest, judging from the description of it by Phillips and Lincoln (1930:88):

> 'Lake Kankakee,' as this river was formerly known, extended approximately from Momence, Illinois, to a point near South Bend, Indiana, an airline distance of about eighty miles in length, but because of the extreme crookedness of the original river channel, it traversed about two hundred and

> forty miles in flowing between the two towns named. The marsh area bordering this stream varied from five to ten miles in width. Previous to drainage these marshes literally swarmed with fur-bearing mammals, ducks and geese.

There are now three flocks of giant Canada geese in Indiana: at the Jasper-Pulaski State Game Preserve there are about 50; at the Tri-county State Fish and Game Preserve, 22; and at Lake Sullivan, Marion County, about 50 (William B. Barnes, *personal communication*).

OHIO — There are free-flying populations of breeding giant Canada geese at three state-owned localities in Ohio: (1) Lake St. Mary's, a reservoir in Mercer County; (2) the Mosquito Creek Wildlife Area in Trumbull County; and (3) the Killdeer Plains Wildlife Area, Wyandotte County. All of these populations originated from private decoy stocks that had been kept for 25 to 30 years by their owners prior to purchase by the state for propagation and release (Karl Bednarik, *personal communication*).

The Lake St. Mary's flock originated with a small nucleus flock released in the early 1940's, and supplemented in 1955 by additional releases. This population, which ranges out for 40 miles from the lake during the nesting season, has remained static at about 1,100 birds despite the fact that 800 to 1,000 are raised each year. Heavy hunting pressure is the chief limiting factor for this flock which is resident in a county having a population of 237 people per square mile.

The Mosquito Creek flock has remained stabilized at approximately 600 geese despite an annual production of about 350 young. This population ranges out for about 15 miles from the refuge area.

The resident flock at the Killdeer Plains Wildlife Area in Wyandotte County has remained static at about 500, and annual production is around 250 young. This flock forages out from the refuge for about 10 miles.

KENTUCKY — The evidence for *maxima* having been the race which nested in Kentucky in former times appears to be secure. Audubon (*in* Ford, 1957:56) described a gander which returned 3 years in succession to nest in a large pond located a few miles from the mouth of the Green River as being "larger than usual" and having underparts of a "rich cream color." The only other large race to which a similar description of a Canada goose in faded breeding condition could apply is *moffitti*.

One of the highlights of my study of museum skins of Canada geese was the "discovery" in the American Museum of Natural History of a large specimen of *maxima*, on which the label read "from the collection of J. J. Audubon" (Table 7). It could have been collected on

Audubon's Missouri River expedition as a taxidermist accompanied him on this trip.

Audubon relates that about 1760, General George Clark found geese abundant the year around along the Ohio River in the Vincennes, Indiana, area and that in 1819, he himself found nests, eggs, and young in the Henderson, Kentucky, area. Surprisingly, as late as 1906, Cooke (p. 72) wrote: "A few breed in Kentucky."

TENNESSEE — The status of the Canada goose as a former nesting species in Tennessee appears to be based on its past occurrence at Reelfoot (Bent, 1925:222). McKinley (1961:5–6) has cited an early comment on the former status of Canada geese at Reelfoot Lake: " 'Geese were very plentiful last fall, and large numbers remain here during the entire year, rearing broods of young. The native goose is much heavier than his migratory brother from Canada [*B. c. interior*] and possessed of an amount of solid sense which, in the human being would lead to fame and fortune' (Kliph, 1881)."

NOTES TO CHAPTER 3

1 The size of the various races of Canada geese appears to be directly related to the severity of the climate on their breeding grounds, to the length of the season for plant growth, and possibly to the period when the ground is free of snow and open water is available. It will be shown in a forthcoming paper that the duration of the molt in adult geese, the length of the incubation period, and the time required for the goslings to reach flight stage are related directly to size, thus offering an apparent basis for a selective effect of climate on size. It will also be shown that *in winter* the distribution of the various-sized races is inversely related to temperature and thereby conforms with Bergmann's law. Recently, Birkebak, Le Febvre, and Raveling (1963, *unpublished*) have evaluated this concept from the standpoint of basal metabolism and heat loss for several races of different size. Their findings offer confirmation of my theory that the cold tolerance of the various races and, consequently, the northward limits of their range in winter are related to body size.

2 References consulted in preparation of Map ii and in related discussions: *Hudson Bay Lowlands* — Geological map of the Dominion of Canada (Map 820A, 1945); *Precambrian Shield and Superior Uplands* — Geological map of the Dominion of Canada (Map 820A, 1945), Fenneman (1931, end map); *Limits of the Wisconsin Glaciation* — Flint (1957), Fenneman (1938, p. 520); *Physiography* — Goode (1943), Raisz (1954), Lobeck (1932); *Prairie biome* — Clements and Shelford (1939), Schantz and Zon (1924); *Deciduous and coniferous forest biomes* — Rowe (1959), Schantz and Zon (1924), Braun (1950), S. Am. For. (1954); *Soils* — Goode (1943), USDA, Yearbook of Agriculture (1938); *Glacial lakes and regions of marine submergence* — Flint (1945, 1957), Flint et al. (1959), Wilson (1958).

3 There is a very thinly scattered population of Canadas (*maxima*) on the western, silt-covered edge of the Precambrian Shield in Manitoba; in Ontario, however, Canadas are virtually absent from the greater part of the Shield.

4 A captive of this flock died and its head was sent to me by Richard W. Vaught. It was identified as *maxima* (Fig. 16).

Migration 4

Molt Migrations

IN ADDITION to migration flights to and from the breeding and wintering areas, the nonbreeding segments of populations of many species of waterfowl in temperate and boreal regions carry out seasonal movements just prior to the molt. Most species of ducks make flights of limited distances, usually to a large lake or marsh, or to the seacoast. On the other hand, the nonbreeding segment of goose populations (the sexually immature yearlings, the unmated adult males, and possibly some adult females whose nests failed during an early stage) may make more protracted flights northward. This is particularly true for the more southerly races of geese. In Europe and Asia, the taxonomic status of the bean geese was confused until recently because of the movement made by the forest race of the bean goose (*Anser fabalis fabalis*) into the Arctic breeding grounds of the tundra race (*Anser fabalis rossicus*) just prior to molting (Johannsen, 1945).

In North America, the longest northward molt flights carried out by waterfowl are made by the nonbreeding components of some populations of giant Canada geese (Map iv). In 1949, Peter Scott and I observed a flock of large, pale Canada geese in the Perry River area (Hanson, Queneau, and Scott, 1956). At that time, it was assumed that these geese were western Canada geese (*moffitti*). In retrospect, I have realized that in all likelihood the geese we observed were giant Canadas. Substantial evidence can now be cited to establish that the great majority of the large, light-colored Canadas seen in the Arctic tundra of the mainland west of Hudson Bay are giant Canada geese. In 1795, Hearne (1958 edition: 285–86) wrote:

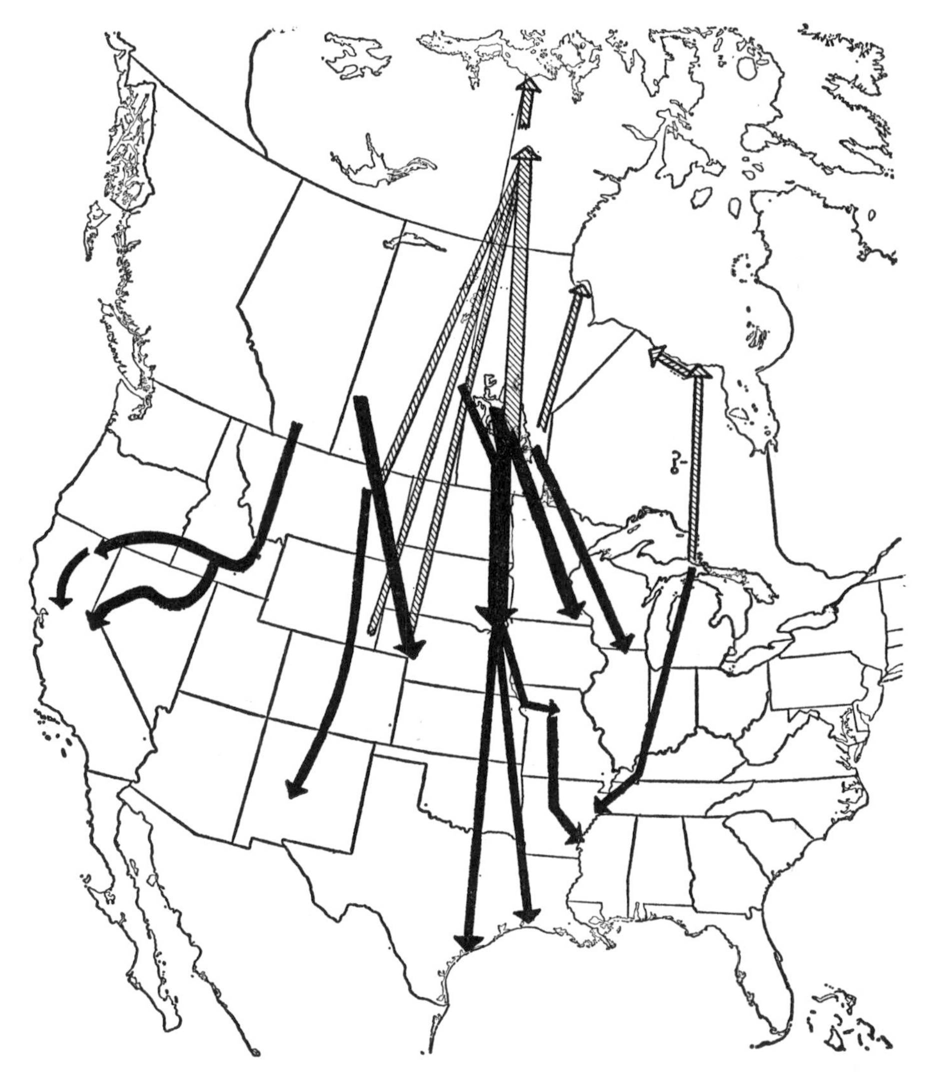

iv. Routes of the molt migration and fall migration of the giant Canada goose.

BARREN GEESE. These are the largest of all the species of Geese that frequent Hudson's Bay, as they frequently weigh 16 or 17 pounds. They differ from the Common Grey [Canada] Goose in nothing but size, and in the head and breast being tinged with a rusty brown. They never make their appearance in the Spring till the greatest part of the other species of Geese are flown Northward to breed, and many of them remain near Churchill River the whole Summer. This large species are generally found to be males, and from the exceeding smallness of their testicles, they are, I suppose, incapable of propograting their species. I believe I can with truth say, that

I was the first European who made that remark, though they had always been distinguished by the name of the Barren Geese; for no other reason than that of their not being known to breed. Their flesh is by no means unpleasant, though always hard and tough; and their plumage is so thick before they begin to moult, that one bird usually produces a pound of fine feathers and down, of a surprising elasticity.

In 1961, Kuyt (1962) observed several hundred large Canadas along the Thelon River, 175 miles west of Baker Lake, Keewatin, some of which were already in molt by June 19. One of these was captured and was found to have been banded at the Bowdoin National Wildlife Refuge in Montana. In 1962, Kuyt captured a flightless Canada goose on the Thelon River (at long 102° 28′ W) that had been banded in Idaho near Homedale (*see* remarks under Idaho, Chapter 3, "Breeding Range"). In 1963, seven additional banded Canada geese were trapped by Kuyt in conjunction with other wildlife studies. These were taken between long 101° 44′ W and 102° 39′ W on the Thelon River and had been banded near or at the following localities: Alderdale, Washington; Soda Springs, Idaho; Redmond Lake, Utah; Bowdoin National Wildlife Refuge, Montana; Lima Reservoir, Montana; Yoder, Wyoming; and Lewellen, Nebraska. The goose banded in Utah and the one banded in Washington were presumably *moffitti*. Two specimens in the molt that were collected for me by Kuyt on the Thelon River in 1964 proved to be typical examples of *maxima:* culmen, 58 and 55 mm; tarsus, 106 and 85 mm; and middle toe and claw, 102 and 87 mm, yearling male and yearling female respectively. (*See* Table 3 for body weights, and Tables 9 and 11 for comparative data on measurements.)

The numbers of large Canadas which spend the summer on the tundra areas of Keewatin are considerable. Dzubin (*letter*, January 16, 1963) writes: "Tom Barry our Western Arctic Ornithologist found numbers of these large Canadas on the Southern Queen Maud Gulf coastline in 1960. Also various counts by U.S.B.S.F. & W. (Wellein & Smith, Chamberlain) and mammalogists of our own Service (Kuyt, McEwen and MacPherson) for the Beverley-Aberdeen-Schultz Lakes–Thelon River complex indicate that from 3,000 to 8,000 large type Canadas moult here every year. Groups vary in size from 20 to well over 500 individuals. . . ."

Conclusive evidence of a molt migration of nonbreeding giant Canada geese to the Barren Grounds was obtained by Tom Sterling in the summer of 1963. Included in a catch of flightless geese, trapped on July 12 on Thelon Island (Fig. 38) in Aberdeen Lake, were 30 banded birds. *Ten of these geese had been banded at Rochester, Minnesota* (one in

1962; nine in 1963)! Three had been banded at the East Meadow Ranch, Oak Point, and most of the remainder were banded at known areas in the wintering range (Lake Andes, South Dakota, and Swan Lake, Missouri). Of particular interest was the recovery of four geese that had been banded at the Hagerman National Wildlife Refuge near Sherman, Texas. Although this locality may be a wintering area of some importance for giant Canada geese, it is significant that it lies directly north (350 miles) of Matagordo Bay, an area of the Texas Gulf Coast also used by wintering giant Canada geese (*see* Chapter 4, "Wintering Grounds").

In the summer of 1964, Sterling and an assistant, aided by three Eskimos, banded 2,300 giant Canada geese on the Thelon River. Included in the catches were over 100 banded geese. The forthcoming report on this field work is certain to be of great interest and value.

The age and sex composition of the 10 recoveries from the Rochester bandings, recorded by Tom Sterling, also bears out Hearne's early findings that the giant Canadas that migrate to the far north are nonbreeders: only 1 of the 10 was an adult female; the others were adult males or sexually immature yearlings.

General observations and sex and age data on the composition of the populations of giant Canadas that remain on the breeding grounds in summer complement findings from Arctic areas. Of the 64 geese trapped in late summer 1962 by George C. Arthur and me at Delta, Oak Point, and Rennie, Manitoba, only 1 (1.6 per cent) was a yearling; yet the following January (1963), yearlings comprised 11 per cent of the Manitoba stocks of *maxima* wintering at Rochester, Minnesota. The previous January (1962) yearlings had comprised approximately 28 per cent of the Rochester flock (Table 25). Earlier, under "Breeding Range," I quoted James L. Nelson regarding the absence of nonbreeding geese in southern Saskatchewan during the molt period. I suspect that most of the nonbreeders from this region molt in the Thelon River area, as one of the banded Canadas trapped on this river by Tom Sterling was from the Wascana flock near Regina. However, Klopman has written (*letter*, May 31, 1962) that the 40 to 60 nonbreeding geese present at Dog Lake during his years of study made only a short molt migration. Their absence from Dog Lake in summer corresponded with the appearance of a similar number of birds at Sleeve Lake, which lies 18 miles directly east of Dog Lake. As there are no geese breeding at Sleeve Lake, it was deemed significant that at the end of the molt period the same number of geese reappeared at Dog Lake.

Catches of flightless geese at the Crex Meadows Wildlife Area in northwestern Wisconsin also indicated that the yearling geese that originate at this locality carry out molt migrations, but the direction or extent of these flights are unknown. Of 46 immatures banded on this marsh in the summer of 1962, only 5 were retrapped the following summer; yet band recoveries indicate that mortality during the ensuing year was not sufficient to account for this seeming drop in numbers (Richard A. Hunt, *letter*, November 21, 1963).

There are two observations which suggest that the southern coastal region of Hudson Bay may also constitute a part of the summering range of nonbreeding giant Canadas although the number that occur eastward into northern Ontario would be very small. W. C. Currie, of the Ontario Department of Lands and Forests, reported to me that several years ago he saw a small flock of Canada geese along the Sutton River in northern Ontario, in which one individual, in particular, was immense and towered above the rest. In the fall of 1961, Fred Close, the manager of the Hudson's Bay Company post at Fort Severn, reported to Harry G. Lumsden that he shot a goose which was very much larger than representatives of the population breeding locally (*interior*). I think it is reasonable to suspect that these large geese were nonbreeding giant Canadas from one of the Michigan flocks, possibly the Seney flock (Map iv).

Autumn Migration Routes

There is a sufficient number of recoveries of banded giant Canada geese to provide some indication of their migration routes in autumn; spring migration routes may differ somewhat, but an adequate number of recoveries is not available to determine the issue either way. The giant Canada geese that nested in the middle and southern portions of their range in the United States have always been relatively sendentary, migrating southward only when forced to do so by the lack of open water. In the Canadian provinces, however, most populations must necessarily move southward in late fall to find open water and food.

One of the characteristics of the giant Canada goose frequently noted by early observers on the Great Plains was its tendency to remain segregated in small flocks, presumably family units. Even during the migration period, this race was not seen in large flocks. George M. Hogue, former Secretary of the North Dakota Game and Fish Board, made the following comment in a letter to the *New York Times* in which he discussed the weights of Canada geese on the prairie (*New York Times*, November 6, 1922):

> One thing I have noticed with reference to these bigger geese is that they seem to travel in small bunches and apparently drift through from the very beginning of the open season, instead of coming through in big flights like the smaller varieties and stopping here for some time to feed up.

The enormous range of *maxima* is matched by the magnitude of its southward dispersal in autumn (Map iv). Banding recoveries of the giant Canada geese nesting on the Seney National Wildlife Refuge revealed that they migrate south through central Indiana, bypass Horseshoe Lake to the east, and winter on Kentucky Lake, areas in western Tennessee, and on the lower White River in the vicinity of Helena, Arkansas (Johnson, 1947).

Recoveries from bandings at the Alf Hole refuge at Rennie, Manitoba, indicated that the bulk of these geese, which gather at this refuge in late summer from the surrounding region, winter on the Rock County prairie of Wisconsin (Map iv).

Band recovery data have shown that the principal breeding grounds of the Rochester flock lie in the interlake country of Manitoba — the long northeast-southwest swath of country lying between the Lake Winnipegosis–Lake Manitoba systems and Lake Winnepeg. They also probably include much of the Saskatchewan River delta, especially the more easterly portions. Included in the catch of giant Canada geese at Rochester in January, 1963, were 15 banded geese, 14 of which had been banded at the East Meadow Ranch, Oak Point, Manitoba. Conversely, the principal area of recovery in Canada of geese banded at Rochester has been the southern third of the interlake area. The recovery pattern to date has been limited to this area because it is relatively accessible to hunters by road, whereas the more northerly sector is wilderness, accessible to hunters only by air or by water routes. Also, few Indians trap or hunt the region.

If a direct line of flight in migration is assumed between the interlake area of Manitoba and Rochester, Minnesota, it will be found to pass through the Agassiz National Wildlife Refuge in northwestern Minnesota and immediately west and south of Brainerd, Minnesota, where a 14-pound goose was bagged a few years ago (*newspaper clipping*, courtesy of Richard W. Vaught). This NW–SE line roughly corresponds to the southern limits of the transition zone between the prairie and the northern pine-hemlock-hardwood forest and has doubtless existed as a flyway route since these zones were established in postglacial times (Map iv). This route may also have been directionally influenced by the drainage pattern of the upper Mississippi River.

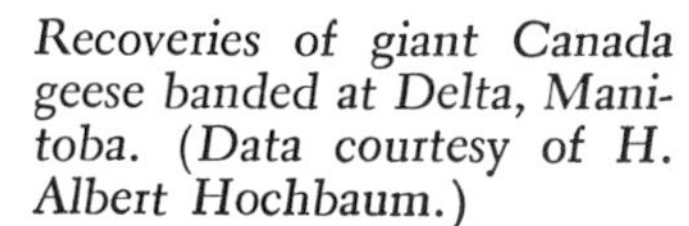
v. *Recoveries of giant Canada geese banded at Delta, Manitoba. (Data courtesy of H. Albert Hochbaum.)*

Recoveries from Delta on the south shore of Lake Manitoba reveal a different pattern of migration (Map v). This population at first migrates directly south to the Missouri River, possibly some flocks via the lower James River valley. Some of these geese apparently then follow the Missouri River to the Swan Lake National Wildlife Refuge (Map v). It should also be noted here that heads of large Canada geese collected for me by Harold Burgess at the Squaw Creek National Wildlife Refuge, Iowa, and by Richard W. Vaught at the Swan Lake refuge, proved to be *maxima* (Figs. 14, 15, and 17).

Another segment of the geese originating from around the south end of Lake Manitoba continues on south to the Gulf Coast of western Louisiana and Texas (Map v). However, had the recovery records from bandings at Delta, Manitoba, not been available, there was other substantial evidence to indicate the presence of giant Canada geese on the Gulf Coast in winter — the following communication by W. P. Reynolds of Olney, Texas, to *Outdoor Life,* May 8, 1922 (*from* Mershon-Holland files):

> Allow me to advise numbers have been killed on the Texas coast that will reach the weight named (16–18 pounds), and some will go higher.
>
> I myself killed a big slate or light grayish goose some years ago near

> Matagorda on the Texas coast that weighed 18 pounds, and tho an unusually large one for that section, parties living there assured me that they had often seen them killed before that large, and a record was retained of one killed there weighing 21 pounds.

More recent confirmation of a movement of some of the giant Canadas to the Gulf Coast of Texas was received in a letter (July 4, 1963) from Rear Admiral Goodall (U.S.N. Ret.) in which he enclosed a clipping of a photograph from the *Houston Chronicle* showing a hunter with a very large goose, definitely a giant Canada. Admiral Goodall also reported that in 1962 there were similar reports of geese weighing between 12 and 14 pounds that were killed in the same area — the vicinity of Aransas National Wildlife Refuge and on the island of Matagorda.

Extensive bandings of Canada geese at the Lake Andes National Wildlife Refuge have resulted in numerous band recoveries along the Saskatchewan River from The Pas area westward into the adjacent portion of Saskatchewan and a scattering of recoveries along the west side of lakes Winnipegosis and Manitoba (Leo Kirsch, *personal communication*). Weights of some Canada geese wintering at Lake Andes are in the category of *maxima*. However, the significance of the recovery pattern of bandings at Lake Andes cannot be fully realized as the banded population in all likelihood included examples of an undescribed race I recently found breeding in the Churchill River Basin (*see* note 1, Chapter 15, "Discussion").

Working westward, the next migration pathway for which there are band recovery data originates in the Kindersley area of western Saskatchewan and terminates at the Crescent Lake Wildlife Refuge in Garden County, Nebraska, and along the North Platte River from Riley to Lewellen, which is a spring-fed portion of the river bordering the southern edge of the Sandhills. The North Platte River is shallow and moderately fast-flowing, with many areas of open water during the winter and numerous islands and bars which afford safe feeding and roosting sites for geese. Bandings of Canada geese nesting in the Cypress Hills of southwestern Saskatchewan have also revealed a migration pathway terminating in the same area of western Nebraska (Harvey W. Miller, *personal communication*). The inquiries of Mershon and Holland had at first suggested to them that the southern limit of the wintering range of the giant Canada geese that nested in the Dakotas — and presumably southern Saskatchewan — was the Platte River in Nebraska, but "Widgeon" (1922), in an account of a hunt on the Arkansas River in central Kansas in the 1880's, described five large Canada geese which clearly were *maxima*, and thereby dissuaded them from this concept.

The giant Canadas that nest on the Bowdoin National Wildlife Refuge in northeastern Montana, the "Hi-Line" geese, take a southward route to their wintering grounds in Colorado and New Mexico that flanks the Rocky Mountain chain (Map iv) (Dale Witt, *letter*, April 19, 1963).

The populations nesting on the prairies of Alberta follow a fairly complex system of routes to their wintering grounds. Recoveries of bandings by Fred Sharp, of Ducks Unlimited, and observations made by Cecil S. Williams have indicated that the giant Canadas banded in the Brooks-Bassano area initially fly more or less directly south. They may follow the Missouri River for part of the way, crossing the Rockies at one of the narrowest points and making juncture with the Snake River near American Falls. The Snake River is then followed west to Twin Falls. At about this point the flight splits, a major segment of it heading south along the water courses to the Humboldt River in Nevada which, in turn, is presumed to guide their route west and south through Nevada to the Carson Sink country. The lakes southwest of this area and the adjoining counties of California, Mono and Inyo, constitute the termini of this route (Map iv).

The second segment (probably much smaller) of the Alberta flight continues westward along the Snake River to the Owyhee Reservoir, just inside the Oregon border, and from there heads southwest across southeastern Oregon to the lake country of northeastern California and, subsequently, during the course of the winter, down the Sacramento and San Joaquin valleys.

A few of the giant Canadas breeding in Alberta must occasionally depart from the Snake River Valley route and fly directly south into Utah. Cecil S. Williams reported (*letter*, October 22, 1963) that in 1939 he shot a Canada goose at Otter Creek Reservoir which weighed 16 pounds and had a 75-inch wingspread. The weight and wingspread of this goose leave no question as to its racial identity.

Prior to seeing recovery data of geese banded in southern Alberta and to having discussions with Cecil S. Williams, I had concluded that information in the Mershon-Holland files afforded nearly incontrovertible evidence that the giant Canadas from the Alberta prairies migrated into northeastern California and south into the Sacramento Valley. Particularly helpful in this respect were the letters of H. L. Betten of Alameda, California. For example, H. L. Betten to W. B. Mershon, January 18, 1931:

> Your letter of the 6th is very interesting and I am glad to know you have acquired so much data in relation to big geese.

There is not the slightest doubt in my mind as to the existence of a species of the white cheeked goose far larger than the authentic Canada goose. In fact, it is a mystery to me how scientists ever overlooked this variant.

The discussion with Ray Holland grew out of my casual assertion that geese weighing as much as 26 pounds had been killed in California. Of course, I regret that this was not the subject for an investigation many years ago when geese were still numerous in the west. However, several of the big geese have been killed in California during the last two seasons and I have been able to secure considerable reliable data. Doubtless more will be forthcoming shortly as I am now in touch with various authorities.

My personal experience with geese covers a period of almost fifty years, and as we formerly had millions of these birds in the state and I was among them on countless occasions, I believe I know something about them. Although I never managed to kill a goose weighing over 20 lbs. I was present when two of the exceptionally large geese were killed — and they were mammoth. As a matter of fact, in the old days honkers weighing up to 16 pounds were quite common and caused little comment. *But they were dwarfs compared with the exceptionally large fellows which were comparatively rare, were never seen in the company of other geese and were confined to small flocks not exceeding a half dozen birds* [Italics by present author].

The following records were reported in a letter of H. L. Betten to W. B. Mershon, February 4, 1931:

Besides those mentioned in the letter to Ray Holland . . . there are records of the following: 20 lb., 2 oz.; 19 lb.; 22 lb.; 8 geese that totalled 111 lbs.; 6 geese that weighed 90 lbs.; and 9 geese that averaged 16 lbs.; one [that weighed] 17 lbs.; information about several weighing in the vicinity of 18 lbs.; a statement by an old market hunter and club keeper (considered reliable) that geese weighing from 20 to 25 lbs. had been killed by him in the Sacramento Valley; and the claims of "Goose" Lewis, a professional goose hunter, that he killed a goose weighing 26 lbs. in 1928 (which I question). The 19 lb. goose was killed two years ago and the 22 and 17 lbs. geese were killed in the season of 1929–1930.

Moffitt (1931*a*:21), in commenting on the weights of Canada geese breeding in California, was also aware that hunters occasionally shot birds of exceptional size:

There has been much speculation and I believe not a little exaggeration on the part of hunters regarding the weights attained by Canada geese. I have personally weighed a number of Canada geese scaling twelve pounds

> and consider eight to ten pounds usual for well conditioned females. [This statement would also hold true for adults of the race *interior* on their arrival at their Hudson Bay nesting grounds in spring. At that time they reach their peak weights (*see* Hanson, 1962*a*).] However, I have heard so many reliable hunters claim to have weighed Canada geese scaling sixteen or even eighteen pounds, I believe it probable that exceptionally large and old males attain this weight.

In his judgment on the influence of age on weight, Moffitt was in error; I have handled hundreds of Canada geese of the race *interior* that have been banded for 10 or more years but failed to note at any time that these individuals were of exceptional weight.

In *Outdoor Life* (April, 1923), the following news item was reprinted: "Yesterday an immense wild goose was killed on the G. Lindauer Centerville ranch. The bird weighed 24 pounds and it was just 7 feet 4 inches from tip to tip of its wings. It is thought the bird is one of the species from Alaska, as Mr. Lindauer declares he has seen thousands of wild geese, but never saw one of this species before. *Alturas* (Calif.) *Plaindealer*."

In reply, the editor suggested that the bird in question might be a giant Canada goose, but authorities he consulted apparently suggested it was a swan. I am of the opinion that this goose may indeed have been a giant Canada goose, as a veteran hunter is not likely to have confused a swan — young or old — with a Canada goose. Doubtless he was confused by the sheer size of it.

There is also basis for believing that a few giant Canada geese cross over the spine of the Rockies into British Columbia in the course of autumn migration. McAtee (1944) called attention to Mayne's (1862: 418) statement: "The Canada Goose is often shot 17 lbs. in weight." Although Mayne was most familiar with the coastal area of British Columbia, he had made one trip into the interior of the province, the area in which the giant Canada would be most likely to occur during molt or fall migration. But we need not depend on a statement from a bygone era to establish the occasional occurrence of *maxima* in British Columbia. Recently, J. Petley, a veteran hunter of Penticton, B.C. (*letter*, September 12, 1962) wrote:

> Several years ago when hunting near Vernon, B.C., I stopped my car beside a small slough which had about two dozen large birds which I took for Canada geese. I was only a couple of hundred yards away from this flock and they appeared so big I could hardly believe they were wild birds and refrained from even trying to move in closer. Markings seemed identical to what we know as a Canada goose but they were almost as big as

> swans. Subsequently they flew and passed directly over me honking like geese but I was still concerned about the breed and merely watched with interest.

In a later letter (December 12, 1962), Mr. Petley wrote that a friend had subsequently reported to him that some years previously he had shot a Canada goose in northern British Columbia that weighed 18 pounds.

Phenology of Migration

SPRING MIGRATION — Most species of migrant geese in North America return to their breeding grounds several weeks in advance of the spring breakup (Figs. 39 and 40). Nest sites are sought out as soon as the thaw begins, and egg laying may be initiated before islands used for nest sites have achieved protective isolation by the loss of ice cover on the surrounding waters.

The spring arrival of giant Canada geese on their breeding grounds in southern Manitoba is well documented. At the Delta marshes, Hochbaum (1955:122) recorded the following pattern of arrival during a 16-year period: March 20–27, 5 years; March 28–April 4, 7 years; and April 5–12, 4 years.

At Dog Lake, Manitoba, 15 years of records have shown that the first arrivals of Canada geese are between March 26 and April 8 (Klopman, *letter,* May 31, 1962). In 1963, the first two pairs of giant Canadas arrived at the provincial refuge, Rennie, Manitoba, on March 26 (H. P. Laws, *department memorandum,* August 24, 1963).

The spring arrival of giant Canadas at the Seney National Wildlife Refuge in the Upper Peninsula is somewhat earlier, usually between March 12 and 18 and not later than March 30 (Glen A. Sherwood, *personal communication*). At the Bowdoin National Wildlife Refuge, Montana, the earliest arrivals in 1963 were seen on March 1; by this date they are often observed on the Missouri River south of Bowdoin. Records kept between 1958 and 1963 have shown that the refuge flock generally arrives during the third week in March (Russell R. Hoffman, *letter,* December 31, 1963).

AUTUMN MIGRATION — The fall migration movements are more variable than the spring flights; in southern Manitoba a large proportion of the local stocks do not leave the breeding areas until forced out by severe weather. In 1963, the Canada geese on the Delta marshes had not left by late November (H. A. Hochbaum, *personal communication*). In 1964, the 800 giant Canada geese at Agassiz National Wildlife Refuge all left on November 19. Their departure coincided with the sudden influx of cold weather and a severe snowstorm (Herbert H. Dill, *personal communication*). The flocks on the refuges in the Dakotas and the one near Ashby, Minnesota, are not reported to leave until well into December. Similarly, Over and Thomas (1921:58) reported that, "In the southern part of South Dakota, the Canada goose remains until January, or until the last 'air holes' freeze over in the Missouri River." Departure time of the flock breeding at the Bowdoin National Wildlife Refuge, Montana, is usually the third week in November (Russell R. Hoffman, *letter*, December 31, 1963). Pirnie (1938) had no band recovery evidence of an exodus of the Kellogg Sanctuary flock in winter, but thought that half of the flock may have moved southward in December and January. In the Mississippi Flyway, the peak movement of giant Canadas, as shown by their arrival at Rochester, Minnesota, and at Rock County, Wisconsin, is at least four weeks behind the main movement of the race *interior* down the more restricted Mississippi Valley Flyway. (*See* Figure 38 in Hanson and Smith, 1950.)

Wintering Grounds 5

THE WIDELY dispersed wintering grounds of the giant Canada goose have been indicated in the previous discussion on migration. Many flocks tend to be relatively nonmigratory if food and some open water are available. It is chiefly the populations nesting in the northern sectors of the Great Plains in the United States and those breeding in the prairie provinces of Canada that are forced to migrate because of the extreme climatic conditions. Temperature per se is seldom the factor inducing a population to migrate; food and water supplies are the overriding factors determining the local distribution of *maxima* on the Great Plains in winter. In primitive times, limitations of both factors on the northern prairies probably resulted in a thinly scattered population with no great concentration of birds being persistently present in any one area.[1] As Irving (1960:355) has stated, "Individual and social behavior must direct animals to localities, and at those seasons, when conditions are well within the physiological tolerance of the population."

Although a few giant Canadas winter on nearly every major wintering ground for waterfowl in the western two-thirds of the United States, it would be pointless to attempt to describe all these areas in detail. Instead, I have deemed it adequate to list the major wintering grounds known to me and to discuss in greater detail areas in Minnesota and Wisconsin with which I have had some personal familiarity.

Canada

ALBERTA — Salt and Wilk (1958:42) have reported that "Small numbers of Canada geese winter regularly on the open waters of the Bow River near Calgary. . . ." The Inglewood Bird Sanctuary, adjoining the

Bow River on the east edge of Calgary, consists of 400 acres and is a wintering ground for several hundred giant Canada geese each year. A spring-fed stream running through the sanctuary and a system of dams and locks on the Bow River help maintain open water the year around.

United States

CALIFORNIA — The evidence for giant Canada geese wintering in California has been discussed in Chapter 4, "Migration." There are three main areas in California in which this race can be expected to be found in winter: the northeast counties, Modoc and Lassen; the lower Sacramento and upper San Joaquin valleys, particularly the various federal refuges; and Mono and Inyo counties in east-central California. The individuals that winter in these areas are derived principally from populations breeding in Alberta.

IDAHO — Canyon County constitutes the wintering area, as well as a breeding ground, for the 6,000 geese in the Homedale flock. In autumn, this flock is augmented somewhat by migrating flocks from southern Alberta en route to their wintering grounds in California.

MONTANA — Dale Witt (*letter*, April 19, 1963) informed me that there is a wintering population of Canada geese on the Missouri River, near Great Falls, that has not been banded. It is logical to expect that a study of these geese will prove that they are *maxima*. This wintering population probably nests in Alberta.

COLORADO — The city park, Bowles Lake, Bass Lake, local reservoirs, and the surrounding countryside constitute both the wintering and the breeding grounds of the Denver flock of 1,000 giant Canadas. In late autumn, this flock is increased by giant Canada geese from the Bowdoin National Wildlife Refuge (Dale Witt, *letter*, April 19, 1963). Bear Creek and the South Platte River are both partly spring fed and therefore afford these geese with open water in winter.

NEW MEXICO — The upper Rio Grande Valley, especially Socorro County, is also a wintering area for giant Canada geese from the Bowdoin National Wildlife Refuge in Montana and for populations breeding in Saskatchewan that are more or less contiguous with the northeast Montana populations. Presumably, the Bosque de Apache National Wildlife Refuge in Socorro County contains most of the giant Canadas that migrate along a route flanking the eastern foothills of the Rocky Mountains in Colorado and the Sangre de Cristo range in New Mexico.

SOUTH DAKOTA — The Lake Andes National Wildlife Refuge and the

Fort Randall Reservoir complex on the Missouri River, in the extreme southeastern part of the state, usually winter about 10,000 Canada geese. The number may vary considerably from year to year. This is a mixed population as, judging from birds collected at Lake Andes that I have seen, probably as many as three races, including an undescribed race, may occur in this area in winter. Weight data furnished to me by Leo M. Kirsch and Charles H. Lacey also leave little doubt in my mind that an important percentage of the wintering flock is *maxima*. However, at least in some years, considerable numbers of *interior* from the Eastern Prairie Flyway may remain much of the winter period in this sector of the state. These wintering flocks obtain water from a flowing well on the refuge and from open portions of the Fort Randall Reservoir on the Missouri River.

NEBRASKA — According to George Schildman (*letter* to David H. Shonk, January 30, 1963), "Nearly all of the Canada geese wintering in the state are on the North Platte river west of McConaughty Dam. This wintering flock varies from a few hundred to 7500 this year [1962–63]. There quite frequently is a small flock wintering on the Platte [River] between Kearney and Lexington and sometimes in the Chapman areas east of Grant Island." Several smaller races of Canada geese stop in Nebraska in migration and some of these may winter in the state (*see* note 1, "Discussion"). However, it is possible that many of the giant Canadas from Saskatchewan may winter in Nebraska in some years. The Crescent Lake National Wildlife Refuge contains 200 to 250 resident Canada geese — probably all giant Canadas — and in autumn this local flock is augmented by about 300 migrants (Harvey W. Miller, *personal communication*). In western Nebraska, the North Platte River is liberally fed by springs draining the Sandhills, thereby keeping it from freezing and making it attractive to wintering geese.

MISSOURI — The Canada goose population wintering at the Swan Lake National Wildlife Refuge has varied in recent years from 50,000 to 100,000. The great majority of this population is *interior* of the Eastern Prairie Flyway, but included in this wintering population is an unknown but possibly an appreciable number of giant Canada geese. The pattern of recoveries of bandings in southern Manitoba and mass bandings at the Swan Lake refuge indicate an important seasonal interchange of giant Canadas between these two areas (Map v, and Richard W. Vaught and Leo M. Kirsch, *unpublished*). It is likely that the giant Canadas are among the last flocks to arrive at the refuge. During late December, 1962, and early January 1963 (the second half of a split hunting season), a number of heads of *maxima* were collected for me at

Swan Lake by Richard W. Vaught and at Squaw Creek National Wildlife Refuge by Harold H. Burgess (*see* note 1, "Discussion").

TEXAS — The principal wintering grounds for giant Canada geese in Texas are on the Gulf Coast, especially in the vicinity of the Aransas National Wildlife Refuge in Aransas, Refugio, and Calhoun counties, and on Matagordo Island.

LOUISIANA — Recoveries from bandings in southern Manitoba (Map iv) reveal that the western coastal parishes, particularly Cameron Parish, are the termini of populations breeding in Manitoba. Probably most of the giant Canadas that winter in Louisiana are to be found on the Sabine National Wildlife Refuge.

LOWER MISSISSIPPI RIVER — The islands and bars of the Mississippi River from the Tennessee-Mississippi state line south to White Castle, Louisiana, were once well-known wintering grounds for Canada geese (Hanson and Smith, 1950), but during the 1920's and 1930's, the number of geese using this vast area of river and adjacent flood plain gradually declined. It has been commonly believed by hunters in states bordering the lower Mississippi River that the development of the Horseshoe Lake Wildlife Refuge in Illinois and its attendant hunting were responsible for the decline of this wintering population — a view with which I have never been fully in accord. The sector of the Mississippi River lying between Cape Girardeau, Missouri, and Cairo, Illinois, was famous before the turn of the century for the number of Canada geese it held in winter. Consequently, it was an important center for market hunting. Recoveries from early bandings at the Jack Miner Bird Sanctuary (Hanson and Smith, 1950:89–91) have clearly shown that this area was the principal terminus for the Mississippi Valley Flyway population; the creation of the Horseshoe Lake Refuge in 1929 merely moved the geese off the river and concentrated them on an old oxbow lake only a few miles from the river. Thus, while other populations of *interior* that by-passed Horseshoe Lake in migration may have contributed importantly to the flocks that wintered along the lower Mississippi River, the evidence indicates that the Mississippi Valley Flyway population of *interior* made only a minor contribution to the total number. On the other hand, there are numerous recoveries from the delta of the White River, from early bandings at the Swan Lake National Wildlife Refuge in north-central Missouri, but the full significance of these recoveries is difficult to evaluate as it is now known that both *interior* and *maxima* were banded at Swan Lake.

During the early period of his quest for information on the giant Canada goose, Ray P. Holland (*letter*, July 13, 1922, to W. B. Mershon) was of the opinion that these geese did not occur on the lower Mississippi River:

> I have never heard of any of them on the lower Mississippi, which is the best place I know of to shoot Canada geese. Anywhere from Cape Girardeau, Missouri, south to Vicksburg [Mississippi], Canada geese [undoubtedly, reference to *interior*] can be found in numbers, but none of the big fellows.

There are, however, data to suggest that fair numbers of giant Canada geese have wintered on the lower Mississippi River in the past and that some still do. Although the race *interior* greatly predominates at Swan Lake in autumn — and presumably accounts for most of the recoveries along the lower White and Mississippi rivers — the mixed racial character of the bandings suggests that at least some of the giant Canadas migrating through the Swan Lake area must have contributed to the populations of Canada geese that have wintered along the lower Mississippi River. In addition to a contingent of giant Canadas on the lower Mississippi River from breeding areas to the northwest, there is also a well-documented flow of giant Canada geese to this area from the northeast. As stated earlier in Chapter 4, "Migration," recoveries of bandings of giant Canada geese at the Seney National Wildlife Refuge have shown that these geese by-pass the Cairo, Illinois, area to the east and winter on Kentucky Lake, areas of western Tennessee, and on the lower White River in the vicinity of Helena, Arkansas.

It must also be remembered that the breeding range of the giant Canada once extended along the Mississippi River as far south as Reelfoot Lake, and that in past times these local resident geese must have attracted and held migrant stocks, thereby helping to establish it as a traditional wintering area.

In retrospect, the remarks of Robert H. Smith, formerly Mississippi Flyway biologist, take on new perspective (Hanson and Smith, 1950: 123):

> The geese using this section of the river [Kentucky-Tennessee line to Baton Rouge, Louisiana] are widely scattered; usually they are in small or medium-sized flocks, but occasionally in large flocks. They show a preference for certain bars, which they use year after year.

The lack of refuge areas and the need to forage widely for food may account for Smith's observations on flock size, but his remarks are suggestive in view of the tendency for *maxima* to associate in segregated family flocks whereas the race *interior* is highly gregarious.

Finally, there are recent as well as older records of very large Canadas killed on the lower Mississippi River. In January 1964, three giant Canada geese were shot near Rosedale, Mississippi, and their skins were preserved. For loan of two of these three specimens and information on local kills along the lower Mississippi, I am indebted to M. G. Vaiden of Rosedale, Mississippi.

The three specimens shot on January 5, 1964, were correctly identified by Vaiden (1964) as *maxima*. The measurements and weights of two of these geese, adult male and adult female respectively, are as follows: wing, 536 and 493 mm; tail, 156 and 149 mm; culmen, 59 and 60 mm; tarsus, 99 and 93 mm; middle toe and claw, 99 and 95 mm; and weight, 12 lb. 8 oz and 9 lb. 7 oz. These specimens are very dark, being similar to the Round Lake, Minnesota, and Denver, Colorado, stocks.

In his letter of March 7, 1964, to me, M. G. Vaiden wrote:

> Mr. George McGee, . . . spent years on the river until a few years ago when he retired. He is now about 83. Mr. McGee tells me that years ago the Canada goose was twice as large as the occasional goose that is now given to him by some hunters for food purposes.

Additional accounts of big geese on the lower Mississippi were supplied by Mr. Vaiden in his letter of March 29, 1964:

> Mr. Dixon Dossett has been to the Horseshoe Lake area twice in the past three years to shoot geese. Mr. Dossett tells me that these geese are large but not like the old "honker" when he was shooting them from the sand bar at the mouth of the Arkansas river that comes into the Mississippi river about 5 miles south of Rosedale, Mississippi. Mr. Dossett is about 55 years old.

Mr. Vaiden also talked to Roberts and Lawrence Wilson of Rosedale, Mississippi, who formerly hunted the river bars near this town. They reported that in November, 1947, they hunted Henrico Bar on the Arkansas side of the river and killed seven geese. Vaiden wrote:

> Of the seven geese taken three were extremely large geese, in fact, at the time they were surely the largest ever taken in this area. One goose, a gander, weighed at Rosedale on grocery balance scales, 15 pounds and 2 ounces; the second weighed, a male, 10 pounds 5 ounces; and the third weighed 8 pounds, 9 ounces, and was a female. . . . Roberts Wilson is 5 feet nine inches, and holding the large one by the bill, it reached from a straight out arm to the ground.

Weights alone are sufficient to identify the first and the third goose as *maxima*, but the weight of the second goose would not exceed an extra-large adult male *interior.* I suspect, however, that it was an immature male giant Canada that was a part of a brood accompanying the adults.

The decline in the population wintering along the lower Mississippi River has roughly corresponded in time to the extirpation of the giant Canada over much of its range; the two events may be linked. A careful study of the racial identities of the Canada geese using this sector of the Mississippi River is now desirable.

MISSISSIPPI — Just before this book was sent to press, I was provided with information which indicates that the reservoirs of northeastern Mississippi, particularly Sardis Reservoir, are important wintering grounds for giant Canadas or have the potential of becoming so. In December 1963, four recoveries were made at Sardis Reservoir of geese banded at the Agassiz National Wildlife Refuge on July 3, 1963. Three of these geese were immature females and one was an adult female. Although all were shot on different days by different hunters, it is possible that two or more were from the same family. However, these recoveries do buttress the concept advanced earlier that many of the giant Canadas wintering in the lower Mississippi River region originate from areas lying to the northwest as well as to the northeast. Further evidence that Sardis Reservoir is used annually by wintering giant Canada geese was received from M. G. Vaiden (*letter*, October 27, 1964). He reported that in 1962 Dr. Y. J. McGaha, of the University of Mississippi, shot a 14 lb. 2 oz goose at this reservoir.

MINNESOTA — The Rochester area, with its combination of prairies and spring-fed streams and rivers that remain open during the winter period, doubtless was a traditional wintering ground for giant Canada geese prior to settlement (Fig. 41). Since then, hunting and the lack of a protected roosting area can be presumed to have held the wintering population at low levels and to have kept it scattered. The rapid increase in the Rochester flock since the early 1950's (Table 16) can be attributed to the presence of Silver Lake. This lake, 20 acres in size, was formed by the damming of the Zumbro River in 1936; water is taken from it by the city power plant for cooling purposes and then returned heated to the lake, thereby keeping a portion of it open throughout the winter and affording roosting sites for the geese (Figs. 42 and 43). The area of open water in winter was greatly enlarged on completion of an addition to the power plant in 1948. A total of 6,600 acres around the lake is closed to hunting, but only part of this area is available to the flock as a large portion of it is occupied by the city of Rochester (Fig. 43). An 800-acre state hospital farm on the outskirts of Rochester, how-

16. Numbers of giant Canada geese at Rochester, Minnesota, at January inventory, 1952–64[1]

Year	Number of geese	Per cent increase over previous year
1952	250	—
1953	325	30.0
1954	650	100.0
1955	800	23.1
1956	1,200	50.0
1957	1,250	4.2
1958	1,300	4.0
1959	1,400	7.7
1960	2,400	71.4
1961	2,700	12.5
1962	4,000	48.1
1963	5,300	32.5
1964	6,000	13.2
Average		33.1

1. State-wide figures for some earlier years are: 1936 — 215; 1939 — 850; 1940 — 2400; 1941 — 587; 1942 — 176; 1945 — 5; 1946 — 100. Most of these wintering geese are presumed to have been *maxima*. January inventories indicate that there were no wild Canada geese wintering in the state in the winters of 1946–47, 1948–49, and 1949–50.

ever, provides the geese with a feeding area where they are relatively undisturbed.

The local history of the flock wintering at Rochester is fragmentary; information available suggests that possibly it has undergone two cycles of increase in recent decades. In 1924, Dr. Charles Mayo, Sr. purchased 15 giant Canada geese in North Dakota for his nearby estate, Mayowood. These captive geese, combined with an artificial feeding program, had, by 1939, succeeded in attracting 500 to 600 wintering geese. By 1940, the year after the feeding program at Mayowood was stopped, there were still several hundred geese in the area in late autumn, but they were probably scattered, and many may have left the area by midwinter. January inventory data for this area between 1940 and 1951 are lacking. Hence, the presumed population increase in the flock for the first few years shown in Table 16 may represent only the rate at which the geese were locally concentrated by the enlarged open area of Silver Lake following the completion of the power plant addition rather than actual change in the population wintering regionally.

An attempt to establish nesting Canada geese at Silver Lake was made as early as 1936, when six geese were purchased. However, limited success was achieved until about 1947 when a flock of 12 large geese from Nebraska was willed to the city by a former patient of the Mayo Clinic who had enjoyed watching the geese at Silver Lake. This pinioned flock was presumably responsible for decoying wild birds to the lake in autumn. The protection this flock subsequently received permitted its rapid buildup. As reported in Chapter 3, "Breeding Range," this increase has not gone unnoticed on their nesting grounds in Manitoba.

WISCONSIN — Two migrant populations of *B. c. maxima* winter in the state: the Rock County flock and the Greenwood Refuge–Mecan Springs flock in Waushara County. In an earlier study (Hanson and Smith, 1950:121), it was presumed that these flocks were comprised of *interior*, but careful study of the Rock County flock on February 4, 1963, by Richard A. Hunt, James B. Hale, Laurence R. Jahn, and me confirmed that the 1962–63 wintering population of 4,000 was comprised almost entirely of giant Canadas (a few *interior* were noted). In recent years, the Rock County flock has tended to remain fairly stable (Table 17).

17. Numbers of giant Canada geese wintering in the state of Wisconsin, 1939–1948[1] and in Rock and Walworth counties, Wisconsin, 1949–64. Data are from January inventories

Year	Number of geese	Year	Number of geese
1939	900	1952	7,300
1940	4,000	1953	3,031
1941	6,000	1954	10,395
1942	4,715	1955	5,019
1943	1,517[2]	1956	4,897
1944	6,350	1957	2,130
1945	4,100	1958	3,570
1946	4,310	1959	3,500
1947	5,000	1960	4,000
1948	5,200	1961	3,588
1949	4,000	1962	2,530
1950	6,921	1963	4,000
1951	3,755	1964	2,100

1. It is presumed that the population of giant Canada geese wintering in Rock and Walworth counties constituted at least 95 per cent of these totals. 2. The severity of weather that winter may have forced the bulk of the population in the Rock-Walworth county area to winter in Illinois.

The prairies of Rock and Walworth counties (and formerly, the Big Foot prairie in McHenry County, Illinois) have been a traditional wintering area for giant Canadas. The Rock County flock now tends to concentrate its feeding in a two-square-mile closed area in Bradford township in the extreme eastern portion of Rock County, but the decisive ecological key to their presence on this prairie area is the Turtle Creek bottomlands (Fig. 44). Prior to freeze-up, lakes Delavan, Koshkonong, and Geneva are used for roosting; in midwinter, the pasture land of the Turtle Creek valley becomes the alternate watering and roosting site for this flock.

The Mecan Springs area in Waushara County has been known for its wintering Canada geese since the time of the first settlers (Fig. 45) (Ralph Hopkins, *personal communication*). From 80 to 100 Canada geese winter annually along the Mecan River and on the adjacent state Greenwood Farms Refuge, but in autumn thousands of geese, principally *interior*, stop to feed in this area. Available information on the local wintering flock leaves little doubt that the subspecies involved is *maxima*. Weights alone are indicative. A 15-pound goose has been reported shot in the area, and many of the geese that stop in the area during the fall are in the 11 to 12-pound class. In January 1963, following a three-week period of extreme cold and deep snow, the Mecan Springs flock left the area and were believed to have resorted to areas along the Wisconsin River. In 1963–64, 450 Canada geese wintered at Mecan Springs and nearby areas.

ILLINOIS — The wintering range of the Rock County, Wisconsin, flock apparently includes adjacent portions of Illinois, as I have had reports that large, light-colored geese are taken each year by hunters in northern McHenry County. In January of some years, a few hundred Canada geese migrate into the Goose Pond area of Putnam County (Frank C. Bellrose, *personal communication*). Bellrose has concluded (correctly, I believe) from his aerial observations of these flights that they were derived from the Rock County flock.

A few giant Canada geese have always wintered in the Horseshoe Lake area and, presumably, at the Crab Orchard and Union County refuges since their establishment. However, I have examined only a few specimens at Horseshoe Lake which I belatedly realized were *maxima:* one, an adult male weighing 11 lb. 13 oz (the heaviest goose I have weighed there); another, an emaciated "crop-bound" male, the head of which, because of its extreme size, was collected and preserved for the skull (Fig. 18). Of the hundreds of thousands of geese observed during the course of 20 years of study in southern Illinois, I have not noted any whose size indicated that it might have weighed over 12 pounds. Nevertheless, banding records suggest that I probably had trapped and banded a few smaller specimens of *maxima:* for example, a recovery from extreme southwestern South Dakota; one from Warren, Manitoba; and one from McClean, Saskatchewan (Hanson and Smith, 1950:88 and 98) (*see* note 1, "Discussion," for a secondary explanation of these recovery records).

An operator of a commercial hunting club near Horseshoe Lake told me that geese weighing between 12 and 15 pounds were occasionally shot there in the early 1930's, but were rarely taken after these years (Fig. 6). This timing corresponds with the phasing out of the great majority of the wild populations. The giant Canada can now, with the increased populations in the north-central states and Canada, be expected to be found with greater frequency in Illinois and elsewhere in the Mississippi Flyway.[2]

Ecological Characteristics of Northern Wintering Areas

Geographical races of species are often, in reality, ecological races (Mayr, 1963). The giant Canada goose is the endemic race of the prairie biome, demonstrably adapted to it, and, when conditions permit, a permanent resident in all but the most northern portions of the eastern tall-grass prairie and the short-grass prairie of the Great Plains. It is the only race which consistently winters in areas where January temperatures average below 32° F. (Table 18). The midsector of this range can be

18. Average January temperatures (F.), snowfall in inches, and duration of snow ground cover at selected areas in the northern portion of the wintering range of the giant Canada goose[1]

Wintering Area	Average January temperature	Approximate normal annual snowfall	Approximate normal number of days with snowcover (1 inch or more)
Calgary, Alberta	13	49	100
Winnipeg, Manitoba	−3	54	120
Great Falls, Montana	24	50	80
Denver, Colorado	32	50	40
Lake Andes, South Dakota	17	30	80
Platte River area, western Nebraska	22–24	25	60
Rochester, Minnesota	17	40	100
Rock-Walworth Counties, Wisconsin	19	50	100
Mecan Spring, Waushara County, Wisconsin	14	50	110

1. Data from Kendrew and Currie (1955), U.S. Dept. of Agric. (1941), and Visher (1954).

considered the "type wintering grounds"; it is of secondary biological importance that some components of the population winter on the Gulf Coast or, as in the case of the Alberta populations, are forced by virtue of geographical circumstances to migrate over the mountains and winter in California. An attempt should be made, therefore, to decipher the salient factors that are common denominators of the northern sector of the wintering range.

The foremost limiting factor that appears to determine the local distribution of flocks of giant Canadas in winter is the availability of open water, usually supplied by spring-fed streams and rivers. Although some refuge and captive populations of this race have demonstrated that they can survive with snow as a source of water, it is probable that wild populations have seldom wintered in areas where they were compelled to use snow for extended periods. As mentioned earlier, the South Platte River at Denver is spring-fed and the North Platte River in Nebraska is liberally fed by springs draining the Sandhills. Farther east, similar conditions also hold: the Zumbro River and its tributaries in the vicinity of Rochester, Minnesota, are heavily spring-fed (Fig. 41); the Turtle Creek bottoms in Rock and Walworth counties, Wisconsin, are largely spring-fed (Fig. 44); and Mecan Springs, Waushara County (Fig. 45), is the second largest spring in Wisconsin. Canada geese have probably wintered in such areas since early postglacial times. Only the Waushara County area lies off the prairie, but the persistence of wintering populations at this traditional site may be linked, in part, to its proximity to former Lake Wisconsin, an area in which the giant Canada may have also nested in early postglacial times.

Coupled with spring-fed waters is the presence of watercress (*Nasturtium officinale*) in the stream beds. In Wisconsin, this plant is heavily used by the wintering populations and it is at least conceivable that it may be a factor in determining the suitability of the environment, particularly in areas where prolonged cold and snow cover (Table 18) would preclude the availability of other green forage.

NOTES TO CHAPTER 5

1 A veteran hunter in South Dakota reported that old Indians on the Rosebud Reservation had informed him that in earlier times their tribesmen made periodic winter trips to the Missouri River, partly for the purpose of hunting deer and partly in order to hunt Canada geese (presumably *maxima*) which provided, in addition to food, feathers for their arrows and tribal regalia (Bert Popowski. 1945. "Wild goose chase." *Field and Stream* 44(10):10–11, 66–69).

2 In the winter of 1964–65, approximately 50 giant Canada geese were trapped at southern Illinois refuges. Measurements of 5 adult males were as follows: Wing, 541.6 (522–553); tail, 162.8 (152–170); culmen, 63.8 (61–68); tarsus, 106.2 (103–110); and toe and claw, 108.0 (98–115). Eleven geese observed in the field by Dennis G. Raveling had been color-marked at Shiawassee National Wildlife Refuge suggesting that the giant Canada geese trapped may have originated from this refuge.

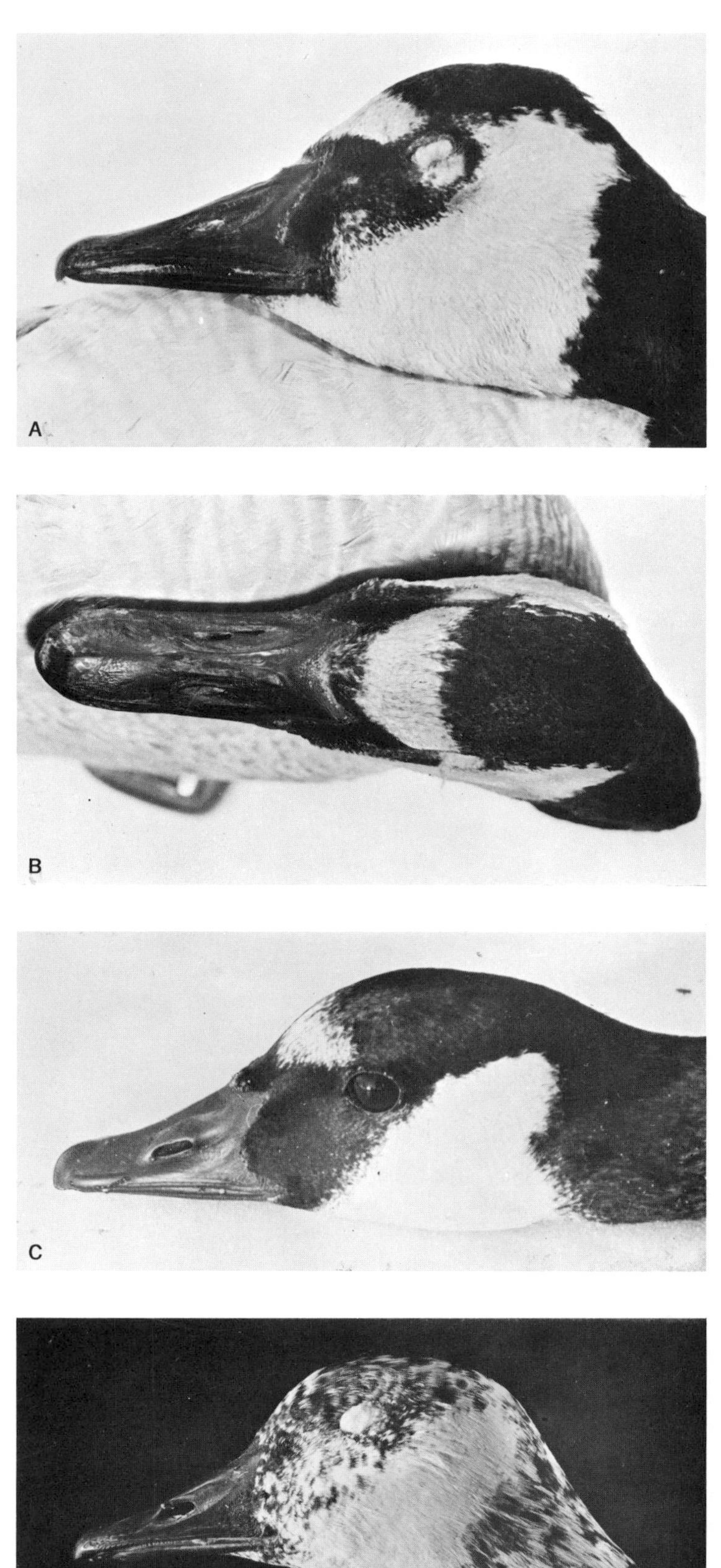

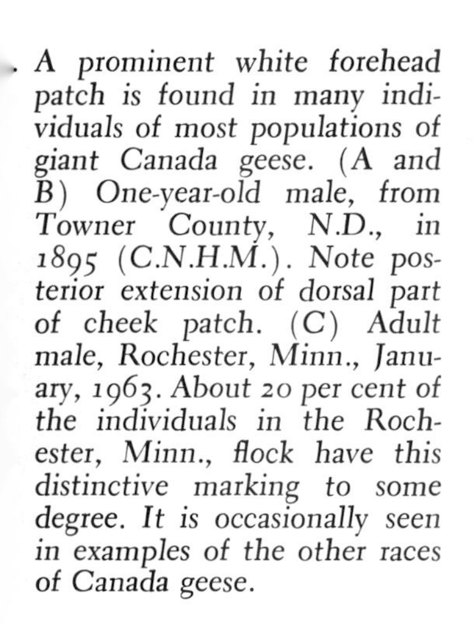
A prominent white forehead patch is found in many individuals of most populations of giant Canada geese. (A and B) One-year-old male, from Towner County, N.D., in 1895 (C.N.H.M.). Note posterior extension of dorsal part of cheek patch. (C) Adult male, Rochester, Minn., January, 1963. About 20 per cent of the individuals in the Rochester, Minn., flock have this distinctive marking to some degree. It is occasionally seen in examples of the other races of Canada geese.

Head of a giant Canada goose, judged to be an adult female, collected September 22, 1944, north of Tilley, Alb. Specimen in the Royal Ontario Museum.

24. A giant Canada goose at Silver Lake, Rochester, Minn., showing extensive white spotting on the neck. Similar examples of localized albinism occur occasionally in nearly all of the races of Branta canadensis. Photograph courtesy of Gordon Yeager.

25. A giant Canada goose trapped in 196 during the flightless period on the Thel River, N.W.T., Can. Note that the whi forehead patch of this goose is so extensi that it is confluent with the white che patch. Two other individuals like this o were also trapped. Photograph courtesy Tom Sterling.

26. (Top and bottom) Giant Canada geese in Rock County, Wis. Note light color of the breasts of these birds which appear pearly gray when in flight in bright sunlight.

27. An adult male giant Canada goose at the Round Lake Waterfowl Station, Round Lake, Minn. Note long wings.

28. (Above) *Giant Canada geese on the property of the Canada Cement Company at Fort Whyte, Man. Note the irregular posterior border of the white cheek patches and also the massiveness of the bills of these geese.*

29. (Below) *Giant Canada geese at the East Meadow Ranch, Oak Point, Man. Note that five of these eight geese exhibit white forehead spots. Compare these geese with brood in Figure 35.*

30. (Above, below, and overleaf, top) *Giant Canada geese at the Inglewood Sanctuary, Calgary, Alb. Note full* U*-bend of neck of bird in foreground in A, white forehead spot and light coloration of bird in B, and body proportions of geese in C and D (overleaf, top). Also note that geese in all photographs have posterior extensions of the cheek patches. Photographs courtesy of Harry G. Lumsden.*

30. (Overleaf)

31. *Examples of giant Canada geese on the Charles M.Russell Game Range, Mont. These geese are transplants from the Bowdoin National Wildlife Refuge. Note massive bodies, long necks, white forehead of the bird on the left, and prominent white neck ring of the bird on the right. Photographs courtesy of Dale Witt.*

32. *A stately and elegant example of a giant Canada goose in the city park in Denver. This goose is a part of the free-flying flock of about 1,000 that range the city park, lakes, and reservoirs of the Denver area.*

33. (Above) *A lake in Kidder County, N.D., which was probably once used by nesting giant Canada geese.* (Below) *A portion of the Waubay National Wildlife Refuge, Waubay, S.D.*

34. *Giant Canada geese at:* (above) *Agassiz National Wildlife Refuge, Minn.;* (middle) *the Lac Qui Parle State Refuge, Minn.;* and (below) *the Carl E. Strutz Game Farm, Jamestown, N.D.*

35. (Above) A brood of nearly full-grown giant Canada geese at the Vorst farm near Lostwood, N.D., in 1914. Note the prominent nail of the bill of bird on the right, and the white forehead spots and the large bills and feet of these geese. Photograph courtesy of Ed Gross and Morris D. Johnson.

36. (Below) A portion of the restored Mud Lake marshes on the Agassiz National Wildlife Refuge, Minn.

37. *Giant Canada geese at* (above) *Agassiz National Wildlife Refuge, Minn. Photograph courtesy of James Thompson; and* (below) *Snake Creek National Wildlife Refuge, N.D. Photograph courtesy of David C. McGlauchlin.*

38. *Part of a catch of flightless giant Canada geese taken with the use of a drive-trap on the Thelon River, N.W.T., Can. Note long necks and massive bills of these geese. The goose just left of center has a "comet-tail" margin of cheek patch which is characteristic of most populations of maxima. Note particularly the auk-like stance of goose at the lower left and compare with bird in Figure 10. Photograph courtesy of Tom Sterling.*

39. Canada geese usually arrive on their breeding grounds prior to the spring breakup. These giant Canada geese are shown on their arrival at the Seney National Wildlife Refuge in the second week of March. Photograph courtesy of C. J. Henry.

40. A mated pair of giant Canada geese at the Agassiz National Wildlife Refuge just after spring breakup. The pair is standing on a disintegrating ice floe. Note the massive proportions of the male on the right and, particularly, its long and massive bill.

41. A tributary of the Root River west of Rochester, Minn. A number of giant Canada geese from the Rochester flock were resting on the creek ice and feeding in exposed areas of adjacent bluegrass pasture when scene was photographed. Photograph courtesy of Robert L. Jessen.

42. Giant Canada geese at Silver Lake, Rochester, Minn. Temperature was 0° F. when photograph was taken.

43. (Top) Giant Canada geese feeding in picked cornfield on the northern outskirts of Rochester, Minn. Temperature was near 0° F. with a 30-mile-per-hour wind when photograph was taken.

44. (Bottom) A portion of Turtle Creek near the eastern border of Rock County, Wis. The extensive bottomland pasture of bluegrass adjacent to the creek is used as a feeding and resting site, as well as for roosting during midwinter. Note goose tracks leading to water's edge.

45. (Above) *Bottomland area adjacent to Mecan Springs, Waushara County, Wis. These bottomlands are heavily used by giant Canada geese throughout the winter. Extensive growth of watercress in this spring-fed stream furnishes the wintering flock with an important source of green food.* (Below) *The Mecan River about a mile from its source. This sector of the river is also used by the flock of giant Canada geese that usually winters in the vicinity of the Greenwood Farms State Refuge.*

46. Horn Island, one of th islands in Dog Lake, Man Nesting is confined to th peripheral areas of thi island (Klopman, 1954)

47. Bluff Harbor in Lake Manitoba. In 1964, seven pairs of giant Canada geese nested on the long, narrow island in the foreground.

48. A giant Canada go nesting on a musk house at Waubay N tional Wildlife Refu S.D., May 1963.

Nesting 6

THE GIANT Canada goose has been the subject of a number of nesting studies although the racial identity of the observed flocks was not known at the time or was wrongly assigned. As a result of the investigations of Kossack (1947 and 1950), Balham (1954), Klopman (1958), Collias and Jahn (1959), and Brakhage (1962), the breeding cycle of *maxima* is probably as well understood as that of any other race. Our knowledge of the nesting of this race has also been enhanced by the observations made by early naturalists and by information gained from management studies.

Nesting Habitat

In its use of such habitats for nesting as cypress swamps, cattail and phragmites marshes, prairie, and muskeg, the giant Canada goose has shown a greater adaptability to a variety of environments than any of the other races of *Branta canadensis*. Perhaps the only common denominators requisite in all of these habitat types are that they are available in large blocks and include bodies of water of moderate to large size, with a depth of at least 30 inches, and preferably containing islands (Figs. 46 and 47).

Williams and Nelson (1943:341) have listed the requisites of good breeding areas for *B. c. moffitti*, based on their studies of this race in the Bear River marshes, Box Elder County, Utah:

> . . . (1) a browsing area available to nesting birds and to paired adults prior to the nesting season; (2) an aquatic feeding area during the brooding period; (3) a brooding environment of open water and resting banks; (4) molting cover (emergents); (5) a browsing area for broods after they are on

the wing (this may be the same as 1); (6) nesting sites isolated from interference; (7) nesting sites providing firm foundations; (8) nesting sites with good to excellent visibility. All these items seem to be essential, or at least very important, in determining breeding populations. Nesting sites are also selected for their proximity to channels and to open ponds that provide avenues to the breeding areas.

Judging from present knowledge, these requisites would hold about equally true for *maxima*. I would only slightly qualify the importance of category 4 — "molting cover." It may at times be important for breeding adults with their young, as Hochbaum (*letter*, May 15, 1962) has reported for the Delta marshes: "Actually the best time to see and examine geese in our study flock is after the flightless season. They practically vanish during the flightless period and it is very difficult to locate the birds in these big phragmites marshes."

Bent (1925:211), quoting Alexander Wetmore on the habits of *moffitti* in the Bear River marshes, Utah, has given a more specific description of the use of tall-grass marshes as areas of seclusion for family groups:

> Here he found numerous places where the thick growth of bullrushes had been beaten down to form roosting places for family parties, well littered with cast-off feathers and other signs of regular occupancy. Here they live in peace and safety while the young are attaining their growth and their parents are molting.

It is doubtful, however, whether phragmites or other dense vegetation is used by family groups except for brooding, resting, or emergency cover. When either feeding or nesting, wild geese prefer to use areas with a clear field of vision. Molting cover would be of little importance to the nonbreeding components of *maxima* (yearlings and unmated adult males) as possibly a major portion of these age-sex classes from Manitoba and Saskatchewan, and possibly from a much larger portion of the range of *maxima*, spend the molt period on the tundra areas of the mainland of Canada, chiefly in the District of Keewatin.

In keeping with the comparative aspect of much of this report, it might be pointed out that the nonbreeding components of the race *interior* in northern Ontario do not seek seclusion in heavy cover during the molt but, when not feeding in the grass and sedge meadows, tend to flock by themselves on the larger lakes of the mainland and on the open water of James Bay off Akimiski Island. However, in considering the aspects of good breeding areas listed by Williams and Nelson, I could not, on reflection, help but be impressed by how well vast areas of the muskeg country of northern Ontario meet these nesting requisites for *interior* in an almost optimal checkerboard arrangement. In contrast, this idealized habitat has been relatively limited over large areas of the Great Plains — especially in recent years because of intensive agricultural practices. Hence, on an areal basis, the potential for very large populations of *maxima* does not exist, as it does for *interior*, in the Mississippi Valley Flyway.

Macoun and Macoun (1909:125) have provided what may be as concise a statement as can be made, even at this late date, regarding the requisite factors in the nesting habitat on the Canadian prairies: "Westward it breeds from Manitoba and the prairie region to the Pacific coast. A few pair breed in almost all the prairie lakes having islands in them and where the waters on the outer fringe of a marsh is over 30 inches deep."

A further insight into the kinds of lakes favored by Canada geese on the Canadian prairies may be found in a comment made by Selwyn (1875), who made many notes on the wildlife in the course of a geological survey from Red River, Manitoba, to Rocky Mountain House in 1873: "ducks of various kinds swarm upon nearly all the lakes and pools and geese are frequently seen, *especially upon the saline lakes*." [Italics by the present author.]

Davie (1898:98) has stated that "Nests of the Canada goose in Dakota are usually situated far away from water on the prairies." Davie's remark probably had its origin in Agersborg's (1885:287) earlier note regarding nesting of Canada geese in Clay, Union, and Yankton counties, South Dakota: "Of the few nests of the Canada goose found, the majority have been far away from any water out on the prairie." It should be noted, however, that Agersborg reported in his introduction that there were no true lakes in these counties, but only a number of reedy swamps. Dzubin (*letter*, January 5, 1963) has also found that in the Kindersley Lake area of Saskatchewan, a few pairs will nest near semipermanent water and then later move the broods as much as five miles to more permanent water areas.

NEST SITES

Perhaps one of the more important factors determining whether prairie marshes and lakes are attractive and suitable to nesting giant Canada geese is the presence of muskrat houses. Everywhere in the prairie sector of its range, nesting on muskrat houses has been characteristic of past as well as present-day populations (Fig. 48). At the Waubay

National Wildlife Refuge, South Dakota, and at the Round Lake Waterfowl Station, Minnesota, free-flying giant Canada geese make extensive use of muskrat houses for nest sites. They are used almost exclusively in the Delta marshes of Lake Manitoba and in the delta of the Saskatchewan River. Most prairie lakes lack islands; the muskrat houses, therefore, offer the ecological equivalent of islets or islands which are so abundant in lakes used by nesting geese in Arctic and subarctic areas. Hammond and Mann (1956), Miller and Collins (1953), and others have likewise stressed the ecological importance of muskrat houses to nesting Canada geese (*moffitti* transplanted to North Dakota in the first instance; native stocks of *moffitti* in California in the second study). Krummes (1941:397) reported that at the Malheur National Wildlife Refuge in Oregon "a close relationship between the increase in muskrat and the number of Canada geese has been noted."

Nesting in North Dakota has been described by Bent (1902:173):

> They nest on islands in the larger lakes and in the sloughs, building two entirely different types of nest in the two localities.
>
> In a large slough in Nelson County on June 2, we found a deserted nest. . . . It was a shallow portion of the slough where the dead flags had been beaten down flat for a space of fifty feet square. The nest was a bulky mass of dead flags, three feet in diameter and but slightly hollowed in the center.

Occasional nesting of Canada geese in trees has been known since the time of Audubon and the early explorers of the Northwest (Coues, 1874). It has been most frequently reported for the race *moffitti*, but the giant Canada was known to have nested freely in trees in northwestern North Dakota (Audubon, 1897), in the Reelfoot Lake area of Tennessee, and in the bootheel area of Missouri (McKinley, 1961). A tree nest of *interior* in the muskeg country of northern Ontario has also been reported to me by an Indian from Winisk. The old nests of ospreys, hawks, herons, or ravens usually constitute the platform for such nesting. Present-day populations of *maxima* in Manitoba are reported to occasionally nest on the tops of hay stacks.

NESTS

Available descriptions of the nests of *maxima* suggest that they are usually considerably larger than either the nests of *interior* (judging from those I have seen in northern Ontario) or the nests of *moffitti* (which Cecil S. Williams and I inspected in the Bear River marshes of Utah and at Blackfoot Reservoir, Idaho). Kossack (1950) reported that the average size of the nests of semicaptive Canada geese (*maxima*) near Barrington, Illinois, was 27 inches by 31 inches. Williams and Nelson (1943) found the average nest of *moffitti* in the Bear River marshes, Utah, to be 25 inches in diameter at its widest point. As geese tend to build their nests with whatever materials are close by, the plant association of a particular habitat would account, in part, for the size of nests (Fig. 49). However, some preference may be exhibited. At Dog Lake, Manitoba, Klopman (1958) found that phragmites was favored for nest construction although prairie grasses were more generally available near the nest sites. The massive, powerful bill of *maxima*, with its prominently serrated edges, may enable this race to more easily handle coarser nesting materials than the smaller races.

Nesting density is discussed in Chapter 12, "Behavior."

PHENOLOGY OF NESTING

In keeping with its wide latitudinal dispersal, the initiation of nesting (the laying of the first egg of a clutch) by various past and present populations of the giant Canada goose have encompassed a period of 5 to 6 weeks, beginning in Kentucky in the Green River region some time in March (Audubon *in* Ford, 1957); at the Trimble Wildlife Area, Missouri, March 20, 1962 (Brahkage, 1962); in the Kankakee marshes of Indiana, between April 15 and May 1 (Butler, 1897); in northeastern Illinois, March 24, 1945 and April 16, 1944 (Kossack, 1950); at Horicon Marsh, east-central Wisconsin, April 4, 1952 (Collias and Jahn, 1959); in Minnesota, as early as the last week in March (Heron Lake, 1894), but generally beginning in the second week of April in the southern half of the state, judging from data in Roberts (1932); in North Dakota (Nelson County), as late as early May (Bent, 1902); and at Dog Lake, Manitoba, April 9, 1955 and April 26, 1954 (Klopman, 1958). The peak hatching period of *maxima* in Manitoba probably averages 3 to 4 weeks ahead of that of *interior* in the Hudson Bay Lowlands. This differential results in a marked difference in the stage of plumage development of the immatures of these two races in winter (*see* Chapter 8, "Characters of Age, Sex, and Sexual Maturity").

Egg Size

The size of the eggs of a number of clutches of *maxima* has been recorded in the literature; other eggs were measured for this study (Table

19). It was reasonable to suspect that *maxima* laid the largest eggs, an assumption which comparative data for the other races support. I have deemed it important to document egg size in the large races as these measurements can be helpful in identifying questionable populations.

INITIATION AND RATE OF EGG LAYING

Kossack (1950) found that the "interval in time between the date a territory was first claimed and guarded, and the date the first egg was laid, varied greatly, but the average time was 13 to 17 days." His observations indicated that egg laying may take place at any time throughout the 24-hour period and that the average rate of laying was 1.5 days per egg. A similar rate of egg laying, 1.6 days per egg, has recently been reported for the giant Canada geese at the Trimble Wildlife Area in Missouri (Brakhage, 1962).

INCUBATION PERIOD

The length of the incubation period of *maxima* has been reported as 26 days by Kossack (1950) and 28.6 days by Collias and Jahn (1959). The difference in the two periods is probably due to somewhat different methods of computing the elapsed incubation time. Kossack (1950:644) stated his method as follows: "Since the incubation usually started with the laying of the last egg in the clutch, the eggs on the following day were then considered as being under incubation for one day. This was the basis on which the incubation period was computed." Collias and Jahn (1959:494) considered that the average length of incubation was "from the date of the last egg until the completion of the hatch."

The incubation is performed by the female and is almost constant. Balham's (1954) intensive observations, during the entire daylight period, revealed that the female leaves the nest only once daily for 4 to 10 minutes, usually between the hours of 5:00 and 7:00 P.M. Collias and Jahn (1959) stated that the female did not leave the nest more than two or three times a day, usually for periods of less than 1 hour.

There are racial differences in the length of the incubation period. Carl E. Strutz (*letter*, June 1, 1964) wrote: "My *maxima* take a day or two longer to hatch than do *interior* or *moffitti*. I find the *moffitti* usually hatch in 27 or 28 days, *interior* takes 28 days and many *maxima* usually take a full 28 and sometimes 30 days for hatching. I have had cackling Canada geese (*B. c. minima*) hatch in 25 or 26 days."

19. Size of eggs in millimeters of *Branta canadensis maxima, B. c. interior* and *B. c. moffitti*

Subspecies and locality	Number of eggs in clutch	Average size	Year collected	Authority	Museum
B. c. maxima					
NORTH DAKOTA					
Unknown	4	95.5 × 62.2	—	Goss (1891)	—
Unknown	3	90.7 × 60.7	—	Bent (1902)	—
Unknown	6	90.0 × 61.3	1897	This paper	AMNH
Unknown	2	86.5 × 59.0	1880	This paper	MPM
Sweetwater Lake, Ramsey Co.	5	91.8 × 60.4	—	This paper	CNHM
Devils Lake, Ramsey Co.	6	93.2 × 58.8	1884	This paper	MMNH
Grand Forks Co.	3	88.4 × 59.8	—	This paper	CNHM
Benson Co.	6	87.0 × 60.2	1901	This paper	AMNH
Jamestown, Stutsman Co.	4	90.0 × 59.3	1880	This paper	AMNH
Rock Lake, Townes Co.	2	88.0 × 58.5	1895	This paper	PM
Benson Co.	6	85.3 × 58.2	—	This paper	PM
SOUTH DAKOTA					
De Smet, Kingsbury Co.	7	89.8 × 61.3	1884	This paper	MPM
MINNESOTA					
Round Lake, Grant Co.	4	89.5 × 62.5	1876	This paper	AMNH
Thief River Indian Reservation	4	91.8 × 59.0	1891	This paper	PM
ILLINOIS					
Hamilton, Hancock Co.	6	87.3 × 60.0	1900	This paper	ISM
Barrington, Cook Co.	Flock average	87.5 × 60.0	1945–46	Kossack (1950)	
COLORADO					
Unknown	—	99.3 × 62.2	—	Sclater (1912)	—
INDIANA					
Unknown	—	90.2 × 57.7	—	Butler (1897)	—
Average		90.1 × 60.1			
B. c. interior					
N. ONTARIO					
Lower Sutton River	3	87.1 × 57.0	1959	This paper	INHS
Lower Sutton River	6	86.1 × 59.7	1959	This paper	INHS
Lower Sutton River	5	86.1 × 58.2	1959	This paper	INHS
Lower Sutton River	6	85.1 × 59.9	1959	This paper	INHS
Lower Sutton River	5	83.6 × 57.3	1959	This paper	INHS
Lower Sutton River	5	82.4 × 57.9	1959	This paper	INHS
Lower Sutton River	5	81.8 × 57.8	1959	This paper	INHS
Average		84.6 × 58.3			
B. c. moffitti					
UTAH					
Box Elder Co.	(174)[1]	87.2 × 59.1	1940's	Williams and Nelson (1943)	
CALIFORNIA					
Unknown	—	86.0 × 58.0	—	Dawson (1923)	—

1. Total eggs measured.

INCUBATION TEMPERATURES

In a series of carefully conducted experiments with giant Canada geese at the Bright Land Farm, Barrington, Illinois, Kossack (1947:126) found that "the average incubation temperature of the embryo is 101.3 F; the average shell temperature 100.4 F; the average breast temperature of the incubating goose 101.1 F; the average maximum temperature reached in the nest 101.1 F. . . ."

Artificial Propagation

We owe the existence of a large portion of our present-day stocks of *maxima* to game breeders and aviculturists throughout the upper Midwest and Great Plains states who, for decades, have propagated giant Canada geese. Usually, little difficulty was experienced in maintaining and breeding this race because it was adapted to the region. Some game breeders achieved notably high success and reproductive efficiency in their management of breeding flocks. Outstanding in this regard is Carl E. Strutz of Jamestown, North Dakota. His procedures are summarized below (from *letter*, August 26, 1963).

Strutz allows pairing to proceed normally while his geese are together in a large flock rather than arbitrarily selecting pairs and then penning them, a procedure which seldom proved successful. Once pairs are established, each pair is placed in a small, grass-sodded enclosure containing a small pond and a nest box 3 feet deep, 2 feet wide, and 3 feet high, the top and back of which are of solid wood. These nest boxes are usually readily accepted, although a few pairs may delay nesting until straw is supplied.

As each egg is layed, Strutz removes it for artificial incubation and replaces it with a wooden egg. When the female begins to pluck her down and brood, the entire clutch of wooden eggs is removed. Strutz reports that he has never had a pair of giant Canadas fail to produce a second clutch — one which they are then permitted to incubate.

Prior to artificial incubation, the eggs of the first clutch are held in a refrigerator at 40 F. Each egg is held slightly tilted with the air-pocket at the top, and every morning and evening they are tilted at opposite angles. If held at 40 F for longer than 7 days, hatching success falls off sharply.

During artificial incubation, Strutz maintains the temperature at 99.75 F and turns the eggs 5 times daily. Because Kossack (1947) found that the incubation period in an incubator having a temperature of 99.5 F at the top was 5 days longer than natural incubation, he suggested that incubator temperatures for Canada goose eggs be held 1.5 to 2.0 F higher. However, the excellent success achieved by Strutz after many years of experience provides assurance that his practices are reliable.

Strutz does not suggest spraying the eggs with water unless the manufacturer of the particular incubator used recommends it. He reports the Humidaire incubator does not require spraying the eggs and that it does an excellent job of incubating goose eggs. Strutz has found that by maintaining the relative humidity between 85 and 92 per cent he has had good results. In 1964, he held the relative humidity of his incubator close to 85 per cent and experienced unusual success.

Growth, Development, and Flightless Periods 7

DATA on growth rates and criteria of age of goslings in the wild represent an important hiatus in the literature. Information of this kind is requisite if brood observations are to be used to full advantage in deducing the phenology and success of a breeding season. To aid in the determination of the age of gosling Canada geese in the field, the series of photographs in Figures 50 and 51 were taken in 1954. The goslings shown, a male and a female, were obtained from the flock at the Bright Land Farm near Barrington, Illinois, and are now realized to have been *maxima.*

The first photograph in the series is of a one-day-old gosling. Subsequent photographs were taken of the goslings at weekly intervals for their first 8 weeks of life; the final photograph shows the stage of development at 10 weeks of age. These young were raised indoors on duck pellets and corn; hence, their development may be at some variance with goslings raised in the wild. In body outline, wild goslings are much more "pot-bellied" because of their intake of bulky forage plants. The series in Figures 50 and 51 are, therefore, presented only as a tentative guide to age until observations and photographs are made of goslings of *maxima* reared under natural conditions. Studies of growth and development in the other races of Canada geese on their breeding grounds are needed. They would be particularly pertinent if carried out in areas where two races of diverse size are present (Baffin Island, Southampton Island, and the Yukon Delta).

Fledging Periods

Balham (1954) found the preflight period of gosling growth in giant Canada geese to be as little as 64 days for one brood and between 70 and 86 days for four others. His studies clearly indicated to him that social facilitation influences the time the first flight is attempted. However, if data are to be useful for comparative studies of species and races of wild geese, they must be based on a standard of comparison — the length of the period of growth required before a gosling becomes first physically capable of flight. Only some kind of frightening action or harassment would probably induce an individual to attempt flight at the earliest age at which it would be capable.

The other races of Canada geese, being smaller, develop more quickly; therefore, they attain flight at an earlier age than the giant Canada. The preflight period for the western Canada goose (*moffitti*) has been given variously as from 49 to 56 days in California (Moffitt, 1931*b*), 63 days in British Columbia (Munro, 1947), and 42 days in Wyoming (Craighead and Craighead, 1949). The latter figure is questionable. No precise data are available for *interior* but, judging from my extensive indirect data from Akimiski Island, the preflight period is approximately 63 days. For the cackling goose (*minima*), the smallest of the races of Canada geese, it is only 42 days (Nelson and Hansen, 1959). The decreasing length of the growing season and the period of open water encountered by the Canada goose as it extended its breeding range northward have resulted in selection for populations able to physically mature in progressively shorter periods of time.

Growth of Primary Feathers and Duration of Flightless Period

The only data available at this time on the growth rate of the primary feathers of the giant Canada goose are given in Graph 1. The lengths of the primary feathers were measured and totaled at approximately weekly intervals; these totals were then computed as a percentage of their final total length. The limited data for these captives maintained on pellets indicate that complete growth of all primaries requires approximately 60 days for immatures and about 70 days for adult males.

Flight is achieved, however, before primary growth is complete. Measurements of the primaries of a series of *interior* shot on Akimiski

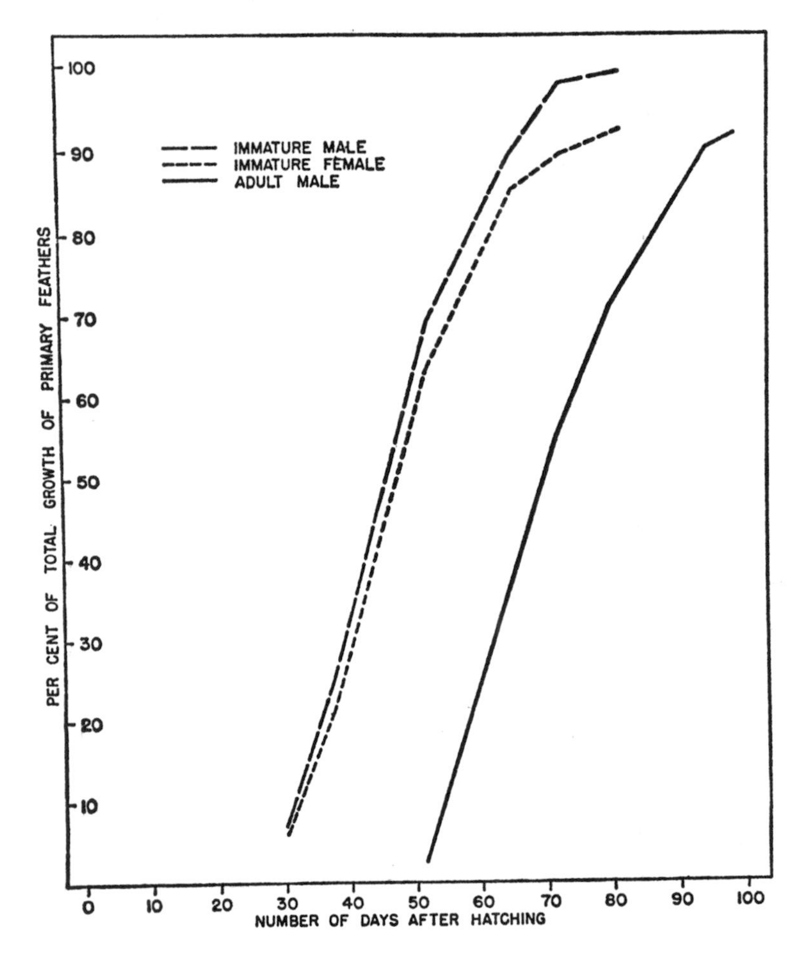

1. *Growth of the primary feathers of two goslings and the parent male. Giant Canada geese obtained from the Bright Land Farm, Barrington, Illinois.*

Island, soon after they were again flying following the molt, indicated that the adults were capable of flight when the total growth of all the primary feathers of the wing was about 85 per cent complete. A large series of measurements of the growth rate of the primaries of captives of this race revealed that the average period between the molt of their primaries and the attainment of 85 per cent of total primary growth is 32 days. A single captive adult male, *maxima*, from the Bright Land Farm, Illinois, required 39 days to reach the 85-per-cent-growth stage (Graph 1). Balham (1954) observed that the time between the molt of the primary feathers and the first flight of adult giant Canadas at Delta and Rennie, Manitoba, was 39 to 40 days, a figure similar to my measurement of a captive adult male. Klopman (*letter*, May 31, 1962) stated that in the Dog Lake, Manitoba, population the molt takes about 40 days.

Differential Timing of the Molt

The period between the initiation of the growth of the primaries of my captive goslings and the onset of the molt and growth of the primaries of the parent male (Graph 1) was found to be relatively long. However, my studies of *interior* in 1958 and 1959 on Akimiski Island in James Bay accounted in part for this unexpectedly large differential; my observations revealed the existence of a sex differential in the timing of the molt in the adults, the molt of the breeding females being found to precede that of their mates by at least from 1 week to 10 days. These findings are at variance with those of Balham (1954), who reported that the onset of the molt in adult giant Canadas occurred 20 to 30 days after the hatching of the young. He also stated that males that had nested molted their primaries earlier than females. It should be pointed out, however, that Balham made these latter observations on only a limited number of birds. On the other hand, I was able to observe and collect many pairs on Akimiski Island.

Characters of Age, Sex, and Sexual Maturity 8

THE BASE upon which the year-to-year management of a wild goose population must rest is a collection of accurate data on the age and sex of individuals in the population. Age and sex data collected during the fall and winter period can be used to make fairly accurate estimates of the productivity of a population in the previous nesting season and the probable number of geese that will be returning to the wintering grounds in the following autumn (Hanson & Smith, 1950:170–71, 192). The validity of these estimates is in large measure dependent on the ability to distinguish yearling adults from older adults as well as from immatures. External morphological criteria useful in aging and sexing Canada geese, *Branta canadensis,* in autumn and winter were summarized earlier (Hanson, 1949). Subsequent studies on both the wintering and breeding grounds have provided additional criteria and established their relative values at various seasons of the year (Hanson, 1962*b*). This chapter, except for a few minor changes, constitutes a reprint of the latter paper.

The criteria of age and sex described here are based on studies of trap and bag samples of *interior* at Horseshoe Lake, Alexander County, Illinois, in the Sutton River area of northern Ontario, and on Akimiski Island in James Bay. The comparative data on *maxima* were obtained at Rochester, Olmsted County, Minnesota.

In the subsequent discussions, the following age classification is used for Canada geese during the fall and winter period: *immatures,* 5 to 8 months of age; *yearlings,* 17 to 20 months of age; *adults,* 29 or more months of age. During the spring and summer, four age classes of Canada geese can be recognized: *goslings, yearlings, 2-year-old adults,* and *old adults* (not all 2-year-old geese can be separated from older adults). Except as noted, the discussions below pertain to sexing and aging geese during the fall and winter period.

Immature Canada geese can be separated from older geese on the basis of plumage characters alone, but for the identification of yearlings and for sex determination, the cloaca must be inspected. Another character useful in age and sex identification is the extensor portion of the carpometacarpus of the wing which, in the male, tends to form a bony spur or knob as the bird ages.

Plumage Characters

The tail feathers, the primary feathers of the wing, and the feathers of the breast and belly are all useful in determining age in Canada geese.

TAIL FEATHERS — The frayed or notched tip of the tail feathers of immature waterfowl, first shown to be a character of age by Beebe and Crandall (1914) and later described by Schiøler (1924) and others, has been widely employed by waterfowl biologists as a criterion of age (Elder, 1946:98; Hanson, 1949:179) (Fig. 52).

In most instances, the tail feathers of the first postnatal plumage may be readily differentiated from those of the adult plumage by their narrower, more tapered outline and their less intense pigmentation. During the winter, the tail feathers of the immature plumage are gradually replaced by the longer, broader, and more deeply pigmented tail feathers characteristic of adult plumage. The tail feathers are generally replaced two at a time, beginning with the central pair. In immature Canada geese in southern Illinois, replacement of tail feathers in winter is largely limited to the central two or three pairs. Observations made in early May, 1959, on the breeding grounds of these geese in northern Ontario, revealed that the tail feather criterion of age could be employed, with nearly comparable accuracy, during the migration period, the last half of April and the first half of May. Most yearlings retained some immature-type tail feathers until the onset of the molt.

The above findings apply particularly to *interior* of the Mississippi Valley Flyway. That they do not apply equally well to all races of Canada geese became apparent in late January, 1962, when 100 individuals of *maxima* were examined by me at their major wintering area, the city park of Rochester, Minnesota, and the nearby surrounding countryside. Plumage characters of the immature *maxima* at Rochester revealed these geese to be in a much more advanced stage of development at a comparable date than the immatures of *interior* in southern Illinois. The tail molt was especially well advanced in the Minnesota birds; at least 5 of the 40 immatures examined had a complete set of adult-type tail feathers. This difference is not surprising, as *maxima* breeding on Dog Lake, Manitoba, build nests (Klopman, 1958:169) at least a month earlier than *interior*, which nests in the Hudson Bay lowlands of northern Ontario and Manitoba. It is advisable to consider both the nesting phenology and the characteristics of the individuals of populations before using collections of tail feathers to obtain age ratios.

PRIMARY FEATHERS — When some of the immature geese in a population have a complete set of adult-type tail feathers in winter, the outermost primary feathers of the wings fortunately provide the investigator with a reliable guide to age determination. The primaries of the immature are pointed; those of the adult are obtuse or rounded (Fig. 53). Cloacal characters in wintering birds provide a nearly infallible guide for distinguishing immature males from adult males; however, the cloacal difference between immature and yearling females is largely limited to the depth of the bursa, which is variable.

The use of the primary feathers of the wing for age determination probably has its greatest potential value in taxonomic studies of museum skins. In making a recent study, in various museums, of the skins of races of Canada geese, I noted specimens that were labeled as to sex but not age or that were incorrectly aged. In these instances a preliminary age determination was made by inspection of the tail feathers and, if these were of the adult type, a conclusive determination was made by inspection of the tips of the primary feathers. Primaries of geese collected in late spring or early summer prior to the molt are usually badly faded by the sun, particularly those of immatures. Being pointed and badly faded, the tips of the primary feathers of the immatures stand out in sharp contrast to the fresh, black tail feathers (Fig. 54) which usually are not attained in full complement until at least midwinter, several months after tail feathers of adults have completed growth.

BREAST AND BELLY FEATHERS — Taverner (1931:31) and Elder (1946:101) have pointed out that breast and belly, or contour, feathers can serve to differentiate immatures from older geese. Breast feathers in the juvenile and postjuvenile plumages of immatures are narrower than in subsequent plumages. They are readily recognized by the prominence of the shaft in each feather, which gives the breast a streaked appearance (Fig. 55). Usually these feathers are replaced during the first winter of life by broader contour feathers, which give the underparts a smoother appearance and a more even coloration. Geese in their second winter of life, and older geese, have wider and somewhat stiffer breast feathers.

Although differences in shape, color, and texture of breast feathers might possibly prove useful in aging geese if no other characters were available, the time required for accurate age determination from these feathers does not make their use an efficient technique.

In late spring and the early part of summer, a sexually mature female that has produced eggs can be distinguished from a sexually immature yearling and from a nonproductive, older adult female by the presence, on the lower breast and belly, of a bare or partially bare area known as an incubation patch (Fig. 56). This area, from which the female has pulled feathers during the incubation period, is subsequently refeathered. By the onset of the wing molt, or shortly thereafter, when the incubation patch has become refeathered, the fresh, unfaded, new feathers stand out in sharp contrast to the worn and faded feathers of the rest of the breast and belly. The patch feathers, therefore, serve to identify a productive female throughout the flightless period in summer. After this period, the remaining old feathers of the underparts of the body are replaced by new feathers, and the feathers of the patch area may become indistinguishable from the rest of the underparts. In a small percentage of females, the patch area produces some white or atypically colored feathers (Fig. 57). These feathers are retained until the next spring and hence, during the winter period, indicate the site of the previous incubation patch (Hanson, 1959:145).

The Wing Spur

The extensor portion of the carpometacarpus bone of the wing can be used in aging geese. In the immatures and yearlings of both sexes in the wintering populations, the skin at this portion of the wing remains feathered (Fig. 58). In adult females, the tip may be partially bare as a consequence of earlier nesting activities. In sexually mature adult males, the tip of the extensor portion of the carpometacarpus is enlarged and sometimes notably knobby, and the skin over it is usually partially denuded of feathers. This condition is a behavioral or anatomical artifact, the result of repeated injury in fights with other adult males. The development of a clublike tip to the extensor serves to increase the effectiveness of the wing as a weapon of defense.

The Cloaca

Within the past quarter century, characters of the cloaca have become widely recognized as criteria of sex and age in waterfowl (Gower, 1939; Hochbaum, 1942; Elder, 1946; Hanson, 1949). A Canada goose can be most easily subdued for examination of the age and sex characters of the cloaca if the goose is held upside down on the operator's lap and the breast of the bird is firmly tucked under the operator's chest. The goose can be readily placed in this position if the operator grasps the bird across the humeri with one hand so that the front of the bird is toward the operator. The bird is in position to be placed on the operator's lap with a half-turn of his hand. If the bird is first held so that it faces away from the operator, it must be turned end over end to be in position for examination. It is awkward to handle a goose in this manner, and the abrupt change in position causes the bird to increase its struggles. The struggles of a bird being held can be greatly minimized if its head and neck are tucked under one wing.

Internal examination of the cloaca with the aid of a nasal speculum (Fig. 59) is relatively simple provided the goose is securely held. A metal rod about 2 mm in diameter and with a smoothly rounded tip should be used as a probe. Extending the cloaca greatly facilitates probing for the bursa. The oviduct attaches to the left wall of the cloaca about 15 mm below the sphincter muscle. To explore the cloaca for an opening to the oviduct, the operator should hold the speculum at right angles to the axis of the bird's body. The site of attachment of the oviduct is most easily inspected if the probe is held nearly parallel to the left wall of the cloaca and moderate lateral pressure is used to stretch and smooth out the tissue in that area. A speculum can be used with no harm to the bird if the cloaca is spread open rapidly and firmly. Repeated, gingerly made attempts, in addition to causing the bird to struggle, are apt to rupture small blood vessels.

THE SPHINCTER MUSCLE — The size, conformation, and color of the external bare area of the sphincter muscle (Fig. 60) provide some immediate clues to the age and sex of the bird being handled. In immatures of both sexes, the sphincter muscle is flat, small in diameter, and pink-red in color. In yearlings, in winter, the bare area is somewhat larger and may be slightly darker in color. In older, sexually mature geese, the sphincter muscle is usually a darker red or varying hues of purple. Adult males can usually be distinguished from adult females by the marked convexity of the muscle in the males. In adult females, the bare area of the muscle is particularly large and flat in cross section (Figs. 60B and D) and the extent to which it can be readily everted usually distinguishes females of this age class from yearling females (Figs. 60A and C).

THE BURSA OF FABRICIUS — Use of the bursa of Fabricius to distinguish immature from older waterfowl has been described, with reference to ducks, by Gower (1939:427) and by Hochbaum (1942:304–6) and, to

Canada geese, by Elder (1946:106–8). Because Canada geese require a longer period to become sexually mature than do most common species of ducks, the bursa of Fabricius is a useful age criterion for distinguishing yearling geese, in winter, from sexually mature adults (Hanson, 1949: 179–80).

The depth of the bursa in immatures ranges between 24 and 35 mm. The average is about 27 mm. In yearlings, it varies between 15 and 24 mm and averages about 17 mm. The bursa is either closed or has been resorbed in approximately 58 per cent of geese in their third winter of life; the remainder may have a remnant of the bursa large enough to be probed, usually to a depth of only 5 to 10 mm. As a criterion of age in Canada geese, the bursa is of value chiefly in separating yearling from older geese, but it must be used in combination with other cloacal characters. Absence of the bursa can be considered quite certain evidence that a goose is 2½ or more years of age.

Closure of the bursal opening into the proctodeum of the cloaca apparently proceeds most rapidly during the second year of life, that is, in the age group achieving sexual maturity in the third spring of life. A female (Fig. 56A) shot along the Sutton River, Ontario, near the coast of Hudson Bay in early June and judged to be nesting for the first time, had a clutch of three eggs, a very restricted incubation patch, and a bursa that had membranous wall that could be probed only with difficulty. She was judged to be nearly 2 years of age.

THE OVIDUCT — In female Canada geese, the membrane that occludes the opening of the oviduct into the cloaca is usually not resorbed until after the geese have left their wintering grounds in their third spring of life. Therefore, determining the presence or absence of an occluding membrane at the juncture of the oviduct with the cloaca is the best way, in winter, to distinguish females approximately 1½ years old from older females. Accurate age determination is of singular importance if the success of the past breeding season is to be estimated from age and sex ratios obtained from trapped birds prior to the hunting season. Studies conducted in preparation for an earlier report (Hanson, 1949:181–82), based on banded birds of known age, indicated there was a high degree of accuracy in differentiating females 1½ years of age from those older. In approximately 97 per cent of the cases, a female goose that, in fall or winter, has adult-type tail feathers, adult-type primary feathers, and a closed oviduct is approximately 1½ years old; if she has an open oviduct, she is about 2½ years of age or older. (In a few individuals, the oviduct may open in the second spring of life.)

THE PENIS — For most species of geese — and especially for the immatures — determining the presence or absence of a penis is essential to accurate sexing. Eversion of the penis from the cloaca of a male bird is a simple matter in ducks. In geese, because of the larger size and greater strength of the sphincter muscle, eversion is more difficult. The technique is the same in both cases — application of a firm downward and outward rolling pressure of the thumbs when placed on either side of the sphincter muscle. The forefingers are used to bend the tail back to aid eversion of the cloaca. At the same time an upward pressure is applied from below the bird by a lateral surface of the middle finger of each hand. Experience in sexing geese and in examining the cloaca with the speculum is probably best gained by working with dead birds.

The presence of an immature penis in a wild goose being examined immediately identifies the bird as an immature male. (Captive geese sometimes develop more slowly than wild geese, and some captive yearling males may have immature penes.) The everted penis of an immature will be seen as a small, corkscrew-shaped organ (about the size of the exposed lead tip of a sharpened pencil) situated in the 8 o'clock position (Fig. 61A).

Present in female geese is a small papilla on the internal edge of the sphincter muscle. This papilla, situated in the 6 o'clock position, is minute in immatures but somewhat larger in older females. It is believed to be homologous to the clitoris in mammals.

The penis of the yearling male goose in winter (Fig. 61B) can be described as intermediate in size between that of an immature and an adult (Fig. 61A and C). In the great majority of instances, penis characters, when considered together with the bursa, are useful in distinguishing yearling adults from older adults. The penis in yearlings is about 4 mm in diameter and 10 mm long. Its most salient features, a pale, translucent color and a smooth surface appearance (Fig. 61B) tend to give it a rather turgid aspect.

The sheathed and unextended penis in older geese, in winter, is generally one and one-half to two times as large as this organ in typical yearlings. It is usually a darker red than the penis in yearlings and is sometimes tinged with purple, particularly toward the distal end. The surface of the sheath is wrinkled in appearance, and the subcutaneous venation is more prominent than that of yearlings.

No discernible change in the development of the penis is seen in wintering immatures prior to northward migration in late February and early March, but in yearlings and adults some enlargement and a tendency to become unsheathed may be noted. By the time Canada geese arrive on the breeding grounds in late April and early May, the penes in all age classes have undergone notable development (Fig. 61D and E) particularly the penis of the immature. In the 2-month interval between departure from the wintering grounds and arrival on the breeding grounds, the penis of the immature (Fig. 61D) undergoes development nearly comparable to that of the adult in spring (Fig. 61E). However, the immature still possesses a large bursa (Fig. 61D) and in most cases immature tail feathers, which provide incontestable evidence that the bird is, indeed, only a 1-year-old.

The maturation of the penis can be observed in a series of photographs (Fig. 62) taken in the course of an experiment on the effects of increased light and of daily injections of gonadotropic hormones on the maturation process.

Discussion and Summary

In addition to the principal characters of age and sex discussed above, there are general clues to the age and sex of the bird. For example, birds in the various age and sex classes vary considerably in size. In winter, the weight of an immature *interior* averages about 1½ pounds less than that of the adult; the weight of a yearling averages one-half pound less. Males are heavier than females by an average of nine-tenths of a pound in the immature and yearling age classes and 1½ pounds in the adult age class. These differentials in weight are reflected in muscular development. With experience, one should have a fairly accurate idea of the age and sex of a live goose simply by grasping it by the humeri. The muscles around the humeri of immatures are not fully developed and they feel stringy; those of older geese, particularly adult males, are well developed. The inexperienced person can most effectively learn to sex and age geese in fall and winter by using the key below.

1. Some or all tail feathers with notched, worn tips and relatively narrow vanes (Fig. 52B, D); color blackish brown — 6
All tail feathers with unnotched, unworn tips and relatively broad vanes (Fig. 52C); color black — 2

2. Primaries pointed at tips (Fig. 53B) — 6
Primaries obtuse or rounded at tips (Fig. 53A) — 3

3. Penis present — 4
Penis absent — 5

4. External portion of sphincter muscle a pale flesh color; penis intermediate in size (Fig. 61B) usually a pale flesh color, translucent and

smooth; bursa open and easily probed, usually to a depth of 15–20 mm; spur of each wing smooth and feathered over at tip *yearling male*
External portion of sphincter muscle dark red or purple; penis large (Figs. 61C and 61A) dark red or purplish in color, with wrinkled surface and fairly prominent venation; bursa closed or, if open, shallow and probed with difficulty; tip of each wing spur enlarged and knobby at tip and more or less denuded of feathers (Fig. 57B) *adult male*

5. External portion of sphincter muscle (Fig. 60A); not much larger than that of immature and light flesh-red in color; oviduct closed at juncture with cloaca; bursa open and easily probed, usually to a depth of 15–20 mm, as in yearling male *yearling female*
External portion of sphincter muscle (Fig. 60B) much larger than that of either immature or yearling female and dark red or blotched with purple; oviduct open and easily probed; bursa closed or, if open, shallow and probed only with difficulty *adult female*

6. Penis present (Fig. 61A) *immature male*
Penis absent *immature female*

On the breeding grounds in spring and summer, the problems of determining the age of geese are more complex and subtle than on the wintering grounds in fall and winter. A male possessing an adult-type tail and a penis that indicates sexual maturity is, nevertheless, only 1 year of age if tips of the primary feathers are pointed, worn, and faded. A year-old female, prior to the molt in her second summer of life, possesses primaries with pointed tips and, with very few exceptions, a closed oviduct. A year-old goose of either sex still retains a large, easily probed bursa.

A 2-year-old goose, after its return to the breeding grounds for the third summer of life, cannot be identified with certainty; in individual cases, however, identification based on a combination of characters may be accurate. An incompletely resorbed bursa together with a normal uninjured wing spur indicates a 2-year-old male. An older male lacks a bursa and, in all probability, has a knobby, enlarged wing spur. A female possessing an open oviduct, a remnant of a bursa, and either a very small or no brood patch may be considered to be 2 years old. An older female, particularly one in the process of egg-laying, has a flaccid, easily distended sphincter muscle, an enlarged oviduct opening, and a prominent brood patch. In a female of either age class, the presence or absence of a brood patch in midsummer should indicate whether or not the individual had attempted nesting in the current season.

Foods and Feeding Habits 9

PRESENT knowledge of the food habits of the giant Canada goose is fragmentary; yet a study of its diet would have several facets of unusual interest. What is the summer and autumn diet of the giant Canada goose in relatively undisturbed original prairie and what are its methods of feeding? Populations of these geese wintering in the northern sector of the range have limited quantities of green foods available because of snow cover. How important are green plants in winter and to what extent does their availability determine the winter range and numbers of these geese? The insights we have on these problems are intriguing. When preparing skins of the Rochester specimens I concluded from the characteristics of their bills, that the giant Canada goose was to a great extent a "seed stripper." This feeding habit had been occasionally observed for *interior* at Horseshoe Lake and at the Horicon refuge. It was, therefore, gratifying that the following report by Robert B. Klopman (*letter*, May 31, 1962) on the giant Canada geese nesting at Dog Lake, Manitoba, confirmed this tentative conclusion:

> Pairs take their young to shore areas to feed almost immediately after hatching. They obviously remain around the nesting site for about a half a day before making the arduous trip to shore. In some cases this means a swim of more than six miles! The surprising thing is that the parents must know the area well enough to know where these feeding localities are. I have seen some of these parties swimming for shore (from great distances I have watched them through glasses), and always they were headed for the few concentrations of bluegrass along or near the shoreline. Although I did not make an extensive examination of the feces of families at this time, I did scour the lake, did know precisely where they were located and

did observe them feeding many times. In 1954 *Poa* was located along the shore in about three broad areas along the north and eastern shores. These were the only areas where I found families. In 1955 I searched for *Poa* growth before the first broods hatched. It was located in about the same places, except instead of being along the shoreline, it was to be found only behind the tree border or the site of the old lake bank. Under these circumstances, the families were not found along the shore, but as much as three to four hundred yards from shore feeding in the areas of *Poa* growth.

Both adults and young strip the seeds from the stalk, and eat only this portion of the plant. This restriction to feeding on *Poa* seeds should not imply a strict preference, however. We all know that goslings can subsist on a variety of succulent green growth, but there simply did not appear to be anything of this sort except *Poa* available at this time.

Subsequently he wrote (*letter*, June 27, 1962):

Let me be specific. Goslings and adults were found to show a marked preference for *Poa pratensis* from about 36 hours after hatching of the young to the beginning of the moult. As one might expect, they ate only the seeds and obtained them by turning the head, placing the stem of the plant in the bill and then running the bill upward so that the seeds fell into the mouth. What is of considerable interest is that parents directed their swim from the nesting islands to areas of bluegrass. A good 80% of the shoreline did not contain any grass or any other food that would have been suitable, particularly for the young. As the parents were never seen to fly to land from the beginning of nesting onward, this knowledge of the location of grass areas must have been passed down as tradition through generations and was not therefore a trial and error search every year. These geese were extremely shy at all times, and in particular avoided islands and shore where tall trees grew close to the water. Of the 100 nests studied only one was found on an island with trees. Yet in the second year the bluegrass did not grow along the shore due to arid conditions, but was found behind the tall trees of the old lake bank. Even under these conditions family parties were found exclusively in the bluegrass areas. This must have been a hazardous undertaking because I observed foxes and coyotes commonly hunting along the old lake bank.

An impressive example of the intensive use of the seed heads of bluegrass by giant Canada geese was seen on the Fort Whyte, Manitoba, property of the Canada Cement Company. Although there were only 22 geese on this area, many acres of bluegrass had been systematically stripped of seeds (Fig. 63). In similar fashion, giant Canada geese at the Alf Hole refuge, Rennie, Manitoba, have been observed to feed on the green seed heads of wild rice (*Zizania aquatica*, Fig. 64). Brakhage

(1962) noted that the giant Canadas nesting on the Trimble Wildlife Area in Missouri commonly stripped the seed heads of bluegrass as well as crabgrass (*Digitaria sanguinalis*).

Waste corn and soybeans are staples in the diet of the giant Canada populations wintering at Rochester, Minnesota (Fig. 43), and at Rock County, Wisconsin. Both flocks also utilize bluegrass which is available to them in pastured bottomlands (Figs. 41 and 44); but when these areas are snow-covered, they seek areas of bluegrass on snow-free hillsides. Laurence R. Jahn (*personal communication*) reported that in Rock County, Wisconsin, acorns (*Quercus macracarpa*), watercress, and skunk cabbage (*Symphocarpus foetidus*) are also eaten. At Mecan Springs in Waushara County, Wisconsin (Fig. 45), the wintering flocks make particularly heavy use of watercress (Ralph Hopkins, *personal communication*).

At Rochester, giant Canada geese will fly out from the lake to feed in periods of severe weather providing the sun is shining. If it is overcast and extremely cold, they are more inclined to roost on the lake most of the day. The geese shown in Figure 43 were foraging for waste corn on a day when the temperature was 0 F and a 30-mile-an-hour wind was blowing. Under these severe conditions they would move only a few steps at a time and then sit down and tuck their feet in their feathers. From this breast-down position, each corn husk was searched for ears missed by pickers or for kernels of corn remaining from earlier foragings.

Endoparasites 10

CENTRAL to the present theme is a comment made earlier by Hanson and Gilford (1961):

> Canada geese of various subspecies range almost the entire continent of North America north of Mexico. Their breeding grounds include, or did include, much of the land lying in the northern half of the continent, aside from most mountainous, heavily forested, and desert regions. The subspecies vary greatly in size, and the ecological conditions under which they breed also vary enormously. It should therefore be apparent from the wide diversity of ranges and habitats that unusual rewards in research are to be gained from studies of the parasites of this group of birds.

The studies I made in collaboration with Norman D. Levine, James R. Gilford, and Virginia Ivens have suggested that certain parasites, particularly the blood parasites and coccidia, may, in effect, be regarded as "biological tracers" that are useful in studying the distribution of the races of Canada geese. They may offer collateral evidence of the degree of isolation of one race from another and, in some cases, the extent to which a race has evolved as a distinctive genetic entity. It is, therefore, instructive to compare available data on the endoparasitic fauna in populations of *maxima* with those reported for the other large races of Canada geese: *interior, moffitti,* and *canadensis.* Some of the investigations of *maxima* have been submerged under other subspecific names. But with our current knowledge of the true identity of various *maxima* populations, these research findings can be related to the correct race and their significance be placed in proper perspective. An error made earlier should also be corrected: Canada geese from Pea Island, North Carolina, are *B. c. canadensis* — not *B. c. interior* as inadvertently indicated by Hanson, Levine, and Ivens (1957).

Protozoa

BLOOD PROTOZOA — A survey of the blood protozoa of *interior* on their wintering grounds at Horseshoe Lake (Levine and Hanson, 1953) revealed the prevalence of three genera in the various age classes as follows: *Leucocytozoon simondi* — immatures 26.6 per cent, yearlings 5.4 per cent, and adults 3.0 per cent; *Haemoproteus* — immatures negative, yearlings 0.9 per cent, and adults 2.4 per cent; and *Plasmodium* (probably *circumflexum*), in only 1 of the 353 geese studied.

In contrast to this high prevalence of blood protozoa in *interior*, blood smears of 227 adult giant Canada geese from southern Manitoba and Saskatchewan produced negative findings (Burgess, 1957). It should be pointed out that Burgess also found the incidence of *Leucocytozoon* in ducks of this region to be extremely low. The cause for this low incidence of these blood parasites in the waterfowl of the prairie was not known to Burgess, but he presumed it involved some subtleties of the host-parasite — vector-environment relationships.

A moderate incidence of blood parasites is indicated for giant Canada geese at the Seney National Wildlife Refuge, Michigan (Wehr and Herman, 1954). Four of the 29 goslings examined were infected with *Leucocytozoon simondi* (14.0 per cent). The habitat of Seney refuge is in a transition area between lake and boreal forests with scattered areas of muskeg-like swamps. The relatively high black fly (*Simulium* sp.) population in this region and in the northern muskeg breeding grounds of *interior*, as compared with most prairie areas, can be assumed to account for the occurrence of *Leucocytozoon* in the populations of geese breeding in these two areas.

COCCIDIA (Eimeridae) — The coccidia of North American geese have been partially surveyed by Hanson, Levine, and Ivens (1957). Some species of coccidia were found widely over the continent in several species and subspecies of geese while others were *seemingly* host-specific or restricted to one or two regions of the continent. Of the five species of coccidia recorded from the four large races of Canada geese (Table 20) only one, *Eimeria truncata*, has been associated with mortality in wild Canada geese (Critcher, 1950; Farr, 1952, 1954). This parasite infected the kidneys of Atlantic Canada geese (*canadensis*) wintering at Pea Island, North Carolina, and was believed responsible for a die-off that occurred there. It was subsequently found in the population (*canadensis*) wintering on the Chesapeake Bay area and in the western Canada goose (*moffitti*) at Tule Lake National Wildlife Refuge, California

20. Species of coccidia (Eimeridae) recorded from *Branta canadensis maxima, B. c. interior, B. c. moffitti,* and *B. c. canadensis*

	Occurrence by subspecies and locality					
				B. c. moffitti		
Species of coccidia	B. c. maxima Seney, Michigan[1, 2, 3, 6]	B. c. interior Horseshoe Lake, Illinois[4, 5]	B. c. canadensis Pea Island, N. Carolina[1, 3]	Malheur, Oregon[4]	Tule Lake, California[4]	Bear River Marshes, Utah[6]
Eimeria fulva	+		+			
Eimeria hermani	+		+			
Eimeria magnalabia[7]	+	+	+			
Eimeria truncata			+		+	
Tyzzeria anseris	+	+	+	+	+	+

1. *From* Farr, 1952. 2. *From* Farr, 1953. 3. *From* Farr, 1954. 4. *From* Hanson, Levine and Ivens (1957). 5. *From* Levine (1952). 6. *From* Wehr and Herman (1954). 7. *E. striata* a synonym.

(Hanson, Levine, and Ivens, 1957). Fortunately, it has not yet been found in geese wintering or breeding in the area between these coastal states. However, much further work needs to be done before all the hosts and the distribution of these parasites are known. For example, a study of the coccidia of the giant Canada goose from the central prairie portions of the range is needed.

Helminths

The helminths of giant Canada geese are known only from the study made by Wehr and Herman (1954*a*) of 29 goslings obtained from the Seney National Wildlife Refuge. Their investigations indicated that at least seven species occurred in this flock (Table 21). In the Canada geese of the Mississippi Valley Flyway, which have been the most intensively studied of any of the races in regard to its parasitic fauna, 12 species were found in 639 geese studied for visceral parasites by Hanson and Gilford (1961) (Table 21). A complete list of the helminths of this race would be considerably larger as some of the parasites found by Hanson and Gilford were too decomposed to be specifically identified and because the liver, kidney, and nonvisceral parts of the body were not examined.

Studies by Wehr and Herman (1964*a*) have indicated that *moffitti* in Utah have few helminth parasites. This also appears to be true for

21. Prevalence of some helminth parasites in *Branta canadensis maxima, B. c. interior, B. c. moffitti* and *B. c. canadensis*

Species of parasite	B. c. maxima Seney, Michigan[1]	B. c. interior Horseshoe Lake, Illinois[3]		B. c. moffitti Bear River Marshes, Utah[1]
	Goslings	Immatures	Adults	Goslings
Blood flukes	3.4	?	?	
Notocotylus sp.		0.3	2.7	
Notocotylus attenuatus	20.7	8.1	7.4	
Prosthogonimus macrorchis		1.0	0.0	
Prosthogonimus sp.	14.0			
Echinostomum revolutum		12.6	9.7	
Zygocotyle lunatum		2.9	7.0	
Hymenolepis sp.	27.5	11.8	6.2	17.6
Drepanidotaenia lanceolata	7.0	16.5	16.3	11.7
Tetrameres sp.	72.4			
Tetrameres fissispina		0.5	0.8	
Amidostemum anseris[6]	35.0[2]	37.5	24.0	41.0
Sarconema eurycerca (adults)		0.0	6.6	
Sarconema eurycerca (filariae)		0.0	5.9[5]	
Ornithofilaria sp. (filariae)		23.8[4]	14.4[5]	
Trichostrongylus tenuis		0.3	1.9	

1. *From* Wehr and Herman (1954): 29 goslings from Michigan and 17 goslings from Utah examined. 2. *From* Herman and Wehr (1954): 329, apparently both immatures and adults, from Pea Island examined; 26 goslings fom Michigan examined. 3. *From* Hanson and Gilford (1961): 639 geese examined. 4. *From* Hanson (1956): 369 geese examined. 5. *From* Hanson, Levine and Kantor (1956). *Sarconema* filariae in 9.5 per cent of yearlings and 4.5 per cent of older adults; *Ornithofilaria* in 21.4 per cent of yearlings and 11.7 per cent of older adults. 6. At Pea Island, North Carolina, 98.0 per cent of 329 *B. c. canadensis* were infected; at Lake Mattamuskeet 49.0 per cent of 106 *B. c. interior* (South Atlantic Flyway geese) were infected (*from* Herman and Wehr, 1954).

blood protozoa and coccidia; no blood protozoa were found by Wehr and Herman, and only a single species of coccidia, the widely occurring *Tyzerria anseris,* was found in this race by Hanson, Levine, and Ivens (1957).

Physiology 11

THE MAJOR events in the life cycle of wild birds — migration, egg-laying, incubation, molting, fall migration, and wintering — are all accompanied by a complex of physiological adjustments. The intricacy of these adjustments, despite years of research by avian biologists, is just beginning to be adequately understood. In 1938, Oscar Riddle, the "dean" of American bird physiologists, impressed with the dynamic seasonal changes that take place in the physiology of pigeons and doves, highlighted decades of study of the physiology of the endocrine organs of these birds in his paper "The Changing Organism." However, most species of pigeons and doves are of tropical and subtropical origin or distribution and are relatively nonmigratory. Also, they do not undergo an annual flightless period as do waterfowl. It is, therefore, not surprising that the evidence now accumulating on Canada geese indicates that more profound annual physiological changes take place in this species than those occurring in the Columbidae. Recent studies of the race *interior* have yielded some findings which I believe are of singular interest. As they undoubtedly apply equally well to most other species and races of migratory wild geese, a summary of the highlights of this research is pertinent to an understanding of the life cycle of the giant Canada goose.

In a study of condition[1] during various events of the life cycle of *interior*, the following parameters received special attention: body weight, weight of muscles of the sternum (*pectoralis, supracoracoideus,* and *coracobrachialis*), weight of the muscles of the tibiotarsus, relative size of the fat deposits, and liver weight. Large variations were found in these components in relation to the seasonal activity of the Canada goose. For a more detailed discussion of the interrelationships of these and other factors than that given below, the reader is referred to Hanson (1962*a*).

Body Weight Relationships

Body weight is, of course, affected by the seasonal shifts in the total mass of the other four components (sternal and leg muscles, liver, and body fat), but during starvation all are reduced simultaneously although not at the same rate. In instances of death occurring from starvation due to impaction of the esophagus from ingestion of rough, dry vegetation or dry soybeans, weight losses amounting to 43 to 46 per cent of average fall weight were incurred. Only in such instances are all fat reserves lost. During such a period of total starvation, fat loss is accompanied by a gradual depletion of protein stores — the muscle tissue itself. This hand-in-hand degradation of fat and muscle tissue during total starvation is presumably due to the need for oxalacetate in the intermediate metabolism of fat. As this component of metabolism is available from glycogen stores for only about 1 day, the oxalacetate needed after this time must be supplied by the glucogenic amino acids. While in theory the oxalacetate required in fat metabolism should be merely recycled and not lost, the conditions of geese in various stages of starvation suggest that there is a considerable "leakage" of amino acids from the degradation of muscle protein in the course of fat metabolism, as well as other metabolic processes directly involving amino acid metabolism and from general wear and tear of protein structures.

In some years at Horseshoe Lake the geese were markedly heavy in mid-November, but with the gradual depletion of food on the refuge and the containment effect of hunting which prevented their foraging outside for food, they underwent a marked reduction in weight by mid-December. Autopsy studies have shown that while they still carried considerable amounts of fat in December, the protein reserves, as indicated by their sternal musculature, had become noticeably reduced. If fat alone could be metabolized to fulfill energy requirements, no serious inroads on the muscle tissue would have been consistently observed in such instances. As Benedict and Lee (1937:92) have remarked in their study of fasting in the domestic goose, "The resistance to fasting is frequently associated with the nutritive condition at the start of the fast. The normal nutritive condition is usually considered to be that obtaining when the animal has not previously been depleted of its protein stores and has a proportional body fat normal for the goose species."

If total protein and fat content of the body are the criteria, there are, in reality, four major extreme nutritive states in Canada geese.

A] *Maximum protein and fat reserves.* Canada geese in this state were found (1) at the end of spring migration, and (2) in caged birds in spring with unlimited food available to them.

B] *Maximum protein and little or no visible fat reserves.* Geese in this state lack any conspicuous fat deposits but have normal development of the body musculature. It is a somewhat transitory state, but is found in yearling geese of *interior* in late May and early June, several weeks after their return to their birthplace in northern Ontario. By the onset of incubation, adult females have lost most of their body fat and are in a similar nutritive state.

C] *Minimum protein reserves and moderate fat reserves.* The atrophied pectoral muscles of a goose in this state classify it as "thin," by common definition, but the individual still possesses a moderate amount of fat. This state is common in wintering geese when, as described above, a food shortage develops or hunting pressure prohibits normal feeding.

D] *Minimum protein and little or no fat reserves.* As also discussed earlier, this state is best exemplified by geese that have died from starvation due to impaction of the esophagus.

I believe that the peak fat reserves carried by the Canada geese of the race *interior* at the time of their arrival on their breeding grounds in northern Ontario represent an evolutionary development which insures survival until well after the occurrence of late snow storms and spring breakup. While in some years the fall migration flight may represent a severe stress for immature geese, my findings indicate that the spring flight is not a stressful event for this population, except for a small percentage of the immature females.

Muscles of Locomotion

Sexual maturation and the various seasonal stresses importantly affect the musculature. The total weight of the sternal muscles in relation to body weight, as shown by data collected in fall and winter, is higher in yearling males than in immature males, and higher in adult males than in yearling males; whereas, proportionate to body weight, the total weight of these muscles in females of all ages is similar to immature males. This progressive increase in the amount of muscle protein in males with age is believed to reflect gradual sexual maturation and the nitrogen-conserving influence of the male sex hormone, androgen.

49. *Female giant Canada goose with newly-hatched goslings at the Seney National Wildlife Refuge, Seney, Mich. Note the light foreheads on the downy young and the absence of eye rings. Photograph courtesy of C. J. Henry.*

50. *Developmental stages of the giant Canada goose as shown by goslings raised in captivity on pelleted food. Goslings are from the Bright Land Farm, Barrington, Ill. Ages are as follows: A = 1 day; B = 1 week; C = 2 weeks; D = 3 weeks; and E = 4 weeks.*

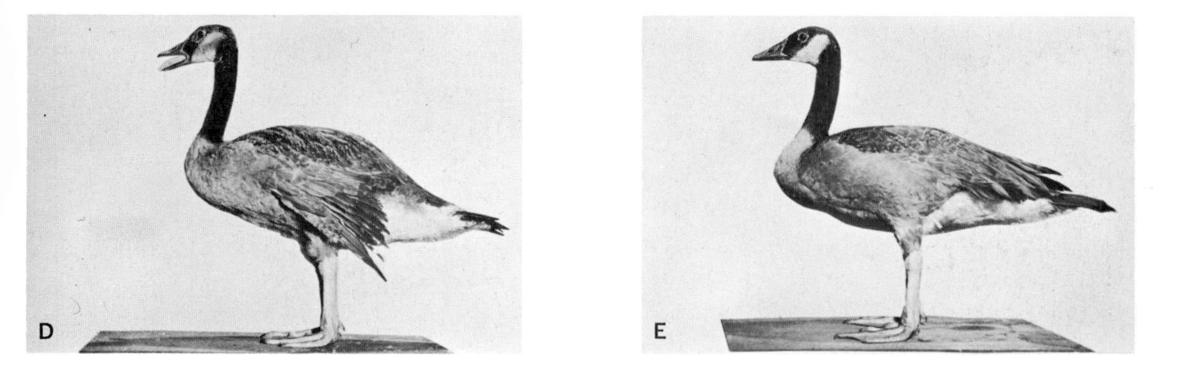

51. Developmental stages of the giant Canada goose continued. Ages are as follows: A = 5 weeks; B = 6 weeks; C = 7 weeks; D = 8 weeks; and E = 10 weeks.

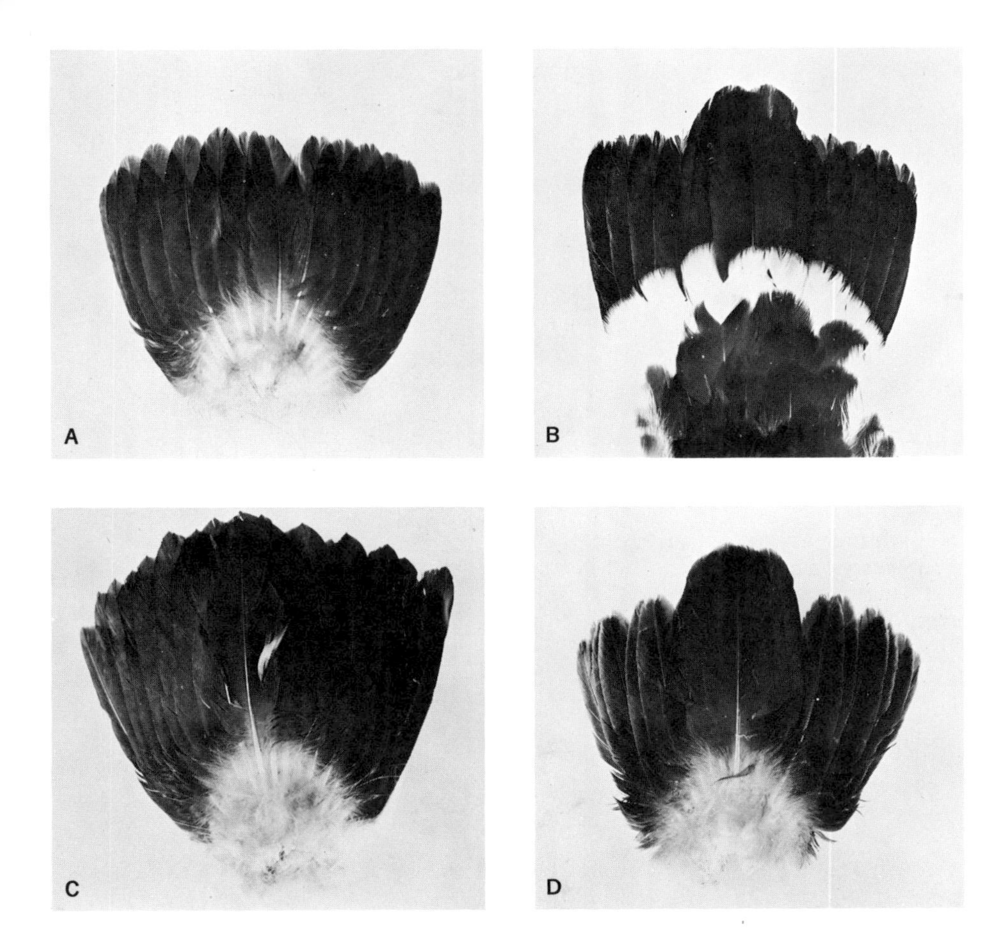

52. Tail feathers of Canada geese (B. c. interior): (A and B) immatures in winter; (C) adult in winter; (D) immature in early May.

53. Primary feathers of the wings of B. c. interior: (A) an adult Canada goose, and (B) an immature, in winter. Feather tips of the adult are rounded; those of the immature are pointed.

54. (At right) Lower dorsal portion of a lesser Canada goose from Perry River, N.W.T., Can., July 3, 1949. Note that faded, primaries of the immature plumage contrast with adult-type tail feathers.

55. (Below) An immature Canada goose (B. c. interior) that still retained the breast feathers of the juvenile plumage in late February, which gives the breast a streaked appearance.

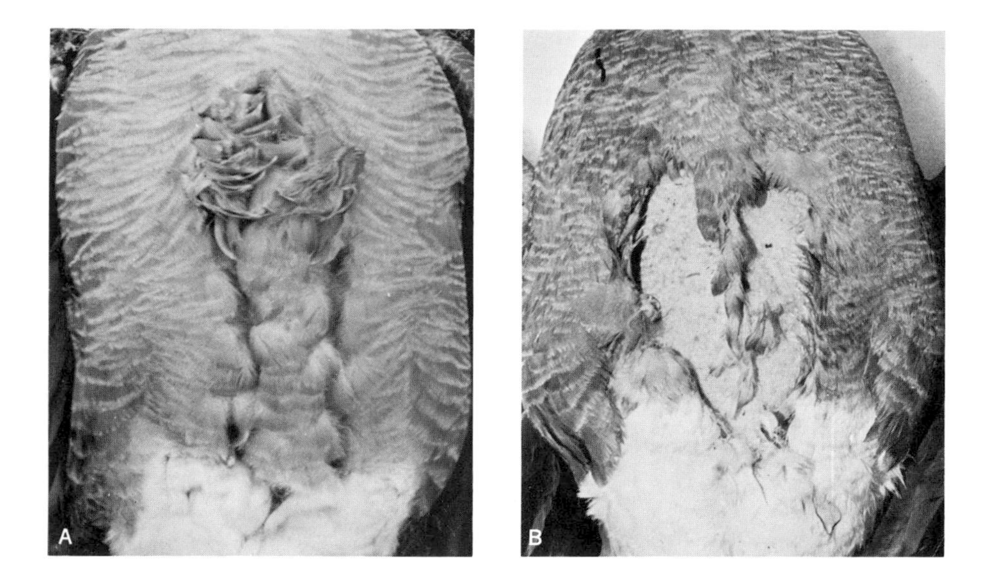

56. Incubation patches of adult female Canada geese (B. c. interior): (A) a 2-year-old that had a clutch of three eggs, June 3; (B) an old adult that had a clutch of seven eggs, May 28.

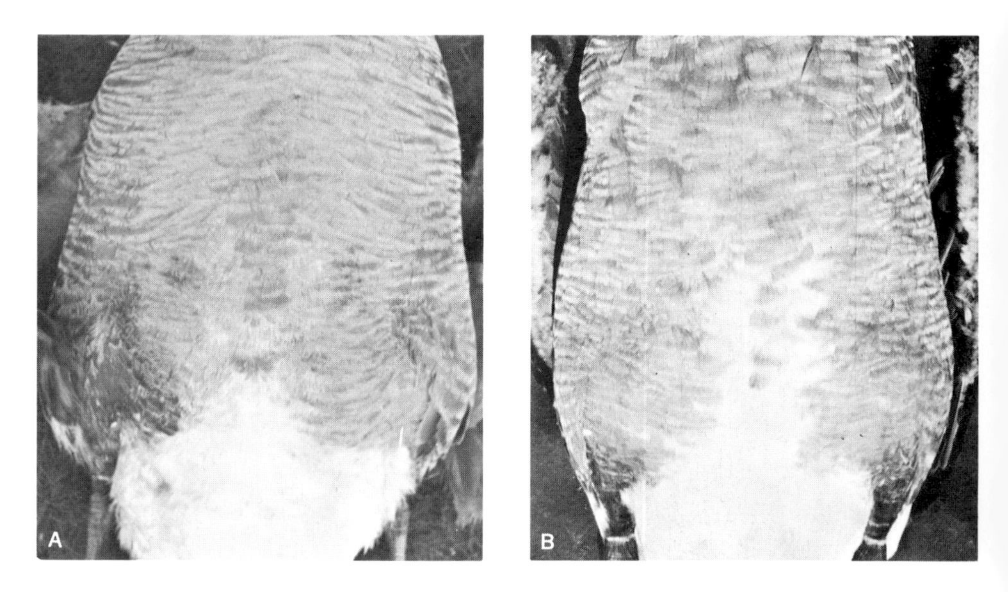

57. Adult female Canada geese (B. c. interior) *in winter:* (A) *with* partially pigmented and (B) *with* unpigmented contour feathers on sites of former incubation patches.

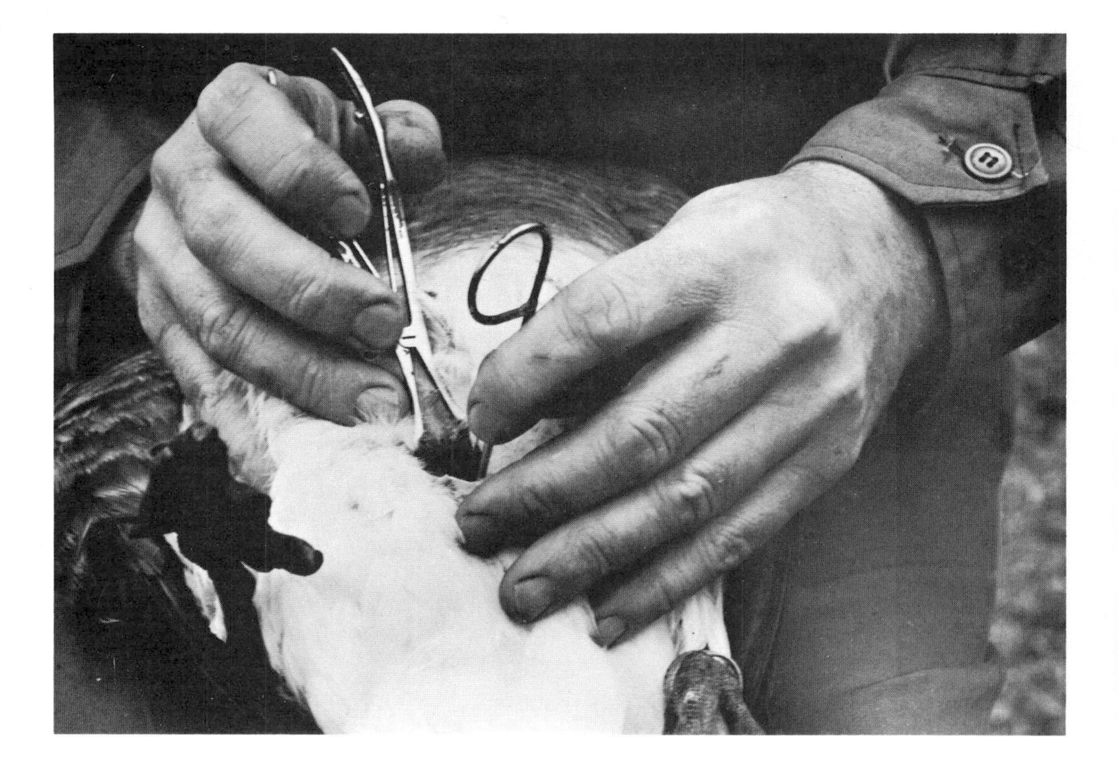

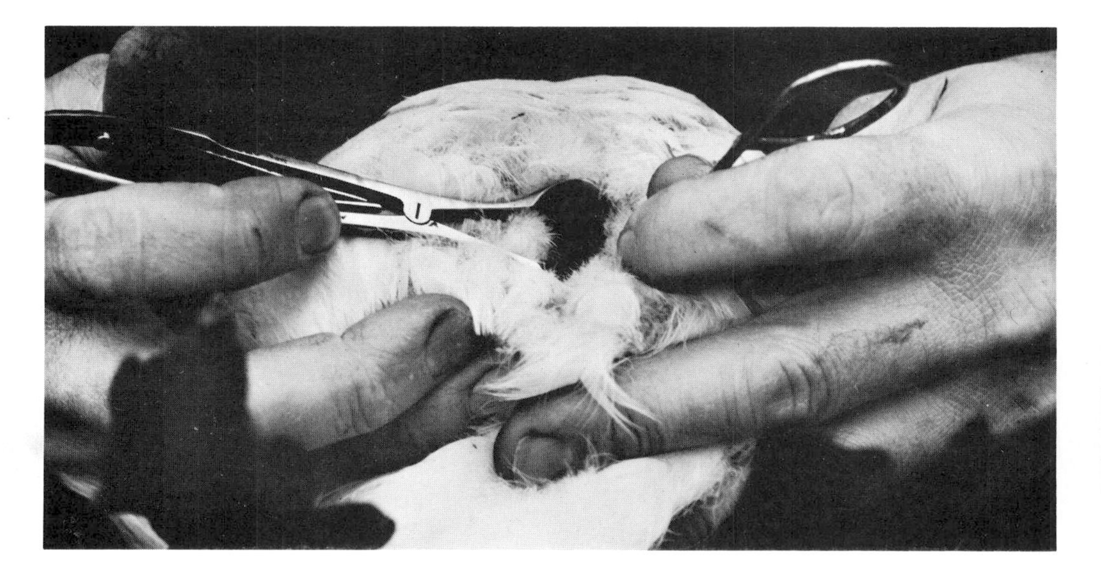

59. (Above) *Use of speculum and probe to explore the proctodeum of the cloaca of a Canada goose for presence or absence of a bursal opening.* (Below) *Method of exploring the cloaca to determine whether or not the oviduct is open at its juncture with the cloaca.*

58. (At left) *Wings of male Canada geese* (B. c. interior): (A) *yearling and* (B) *adult. The enlarged, knob-like portion of the carpometacarpus of an adult is usually conspicuous.*

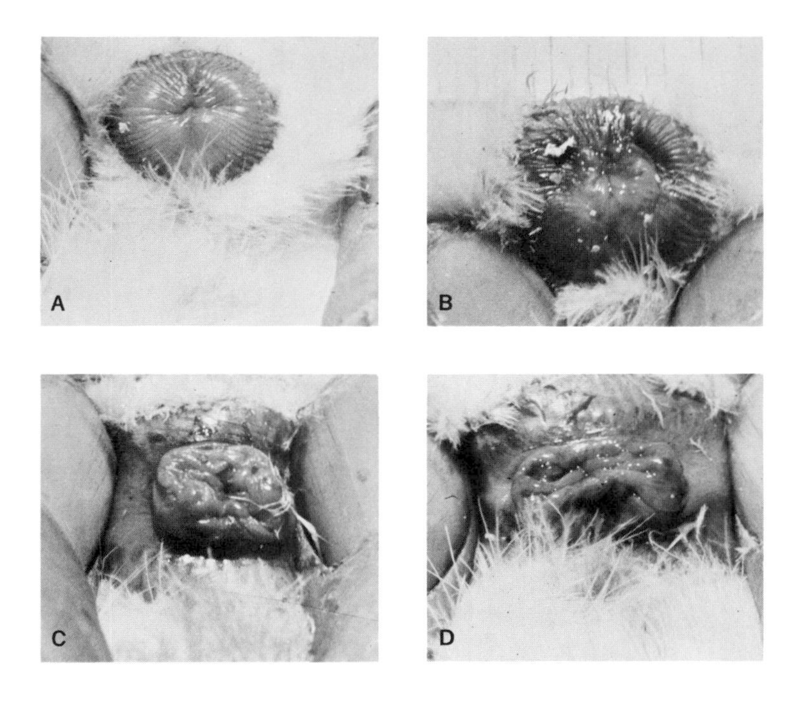

60. Anal sphincter muscles of female Canada geese (B. c. interior) in winter: (A and C) immature, and (B and D) adult; (A and B) the muscle in repose, and (C and D) the muscle partially everted.

61. Penis of B. c. interior: (A) an immature during the first winter of life; (B) a yearling during its second winter of life; and (C) an older adult in winter. The stage of development of the penis of (D) a yearling during its second spring of life (an immature the previous winter) is nearly comparable to that attained by (E) an adult in spring.

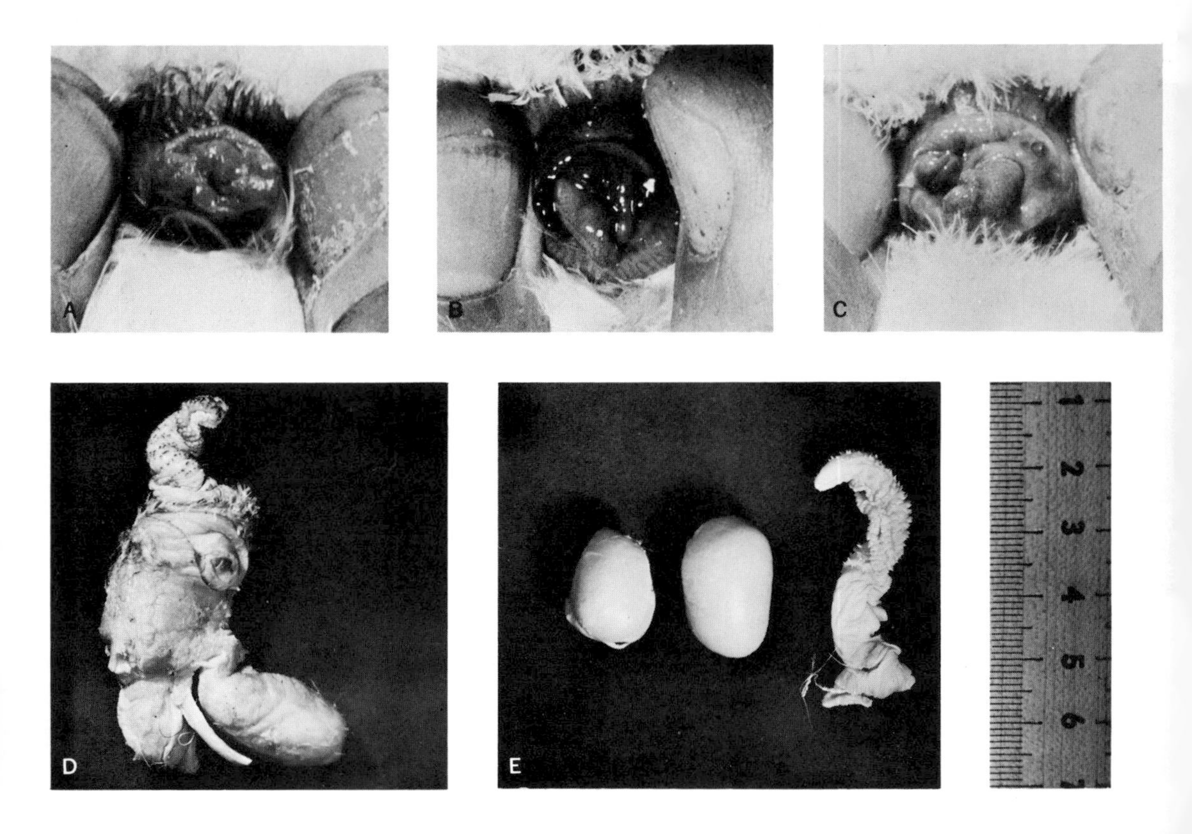

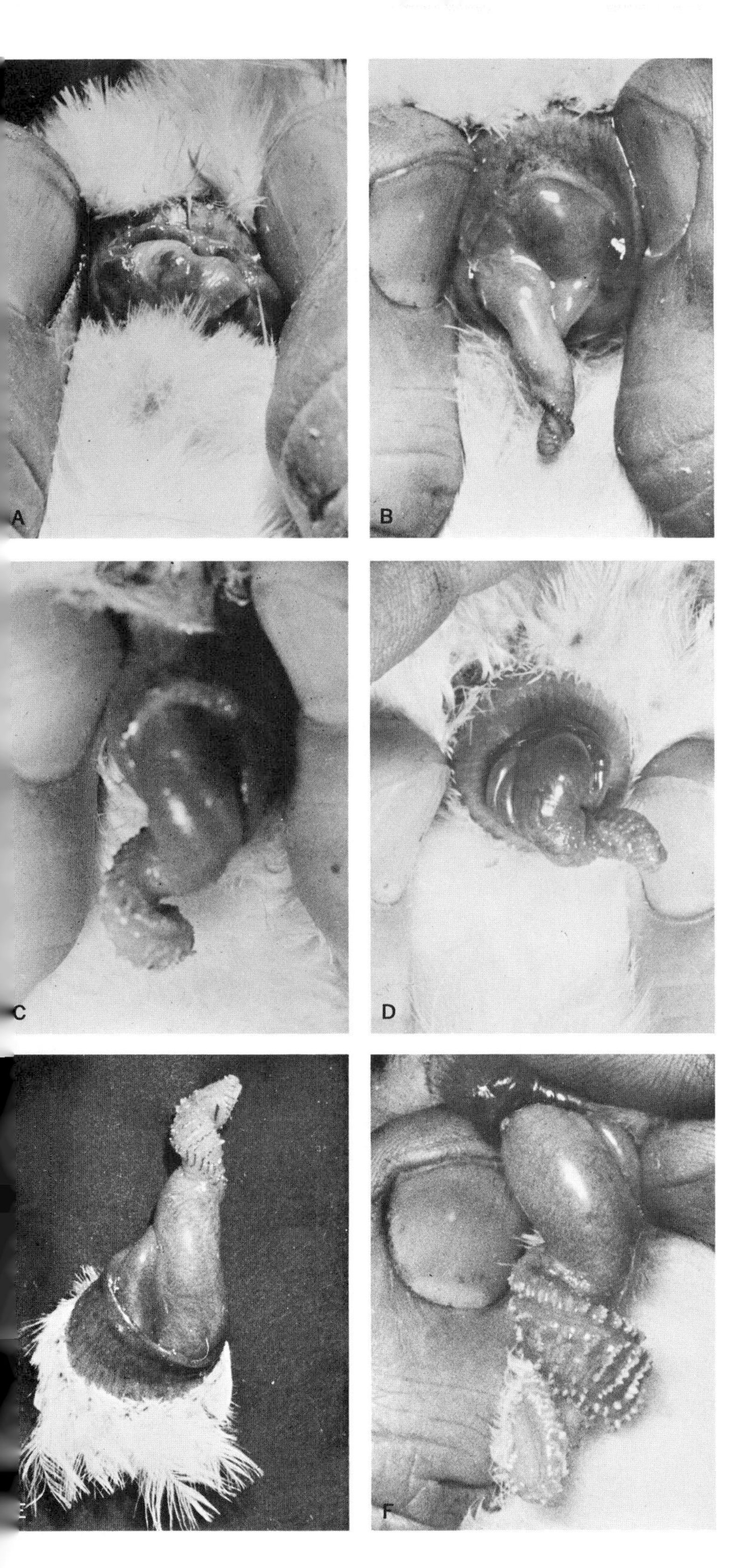

62. (Overleaf)

63. *Bluegrass with seed heads (*Poa*) stripped by giant Canada geese. Photograph taken on the property of the Canada Cement Company at Fort Whyte, Man.*

62. (Overleaf) *Stages of development of the penis of Canada geese (*B. c. interior*) as shown by an experiment involving use of light and hormone: (*A*) adult male, untreated in early February; (*B*) yearling male, March 2, after receiving 16 hours of light a day for 18 days; (*C*) captive yearling male, April 11, caged outdoors; (*D*) yearling male that received gonadotropic hormone daily and 16 hours of light for 4 days; (*E*) stage intermediate between D and F; (*F*) yearling male, March 2, after being given 16 hours of light and gonadotropic hormone injections daily for 18 days.*

64. (Above and below) *Giant Canada geese stripping the green seed heads of wild rice at Rennie, Man. Plumage characteristics and size indicate that the geese in the lower picture are, from left to right, an adult female, an immature, and an adult male.*

65. (Above) Aggressive display at nest site by a male giant Canada goose in response to disturbance by man. Scene is in the Delta marsh, Man. Photograph courtesy of Roger W. Balham.

66. (Below) A pair of giant Canada geese scouting out a nesting territory at the Agassiz National Wildlife Refuge. Photograph courtesy of James M. Thompson.

67. Mutual display by a pair of giant Canada geese in the Delta marsh, Man. Mutual posturing accompanied by frequent calling is usually an aftermath of an aggressive display or an attack by one gander toward another. Photograph courtesy of Roger W. Balham.

68. An entire family may exhibit the sigmoid neck posture as shown here when the gander threatens the males of other families. Photograph courtesy of Roger W. Balham.

69. *A male giant Canada goose in an aggressive posture. This male and its mate are shown at their nest site in the Agassiz National Wildlife Refuge about 3 days prior to hatching of the eggs. Photograph courtesy of James M. Thompson.*

70. *A nesting platform in use at the Waubay National Wildlife Refuge, May 1963. Photograph courtesy of Robert R. Johnson.*

71. (Above) *Type of nesting platform used at* East Meadow Ranch, Oak Point, Man.

72. Washtubs have been used at the Trimble Wildlife Area, Mo., for nest sites in establishing a breeding flock of giant Canada geese. *Photograph courtesy of Don Wolldridge.*

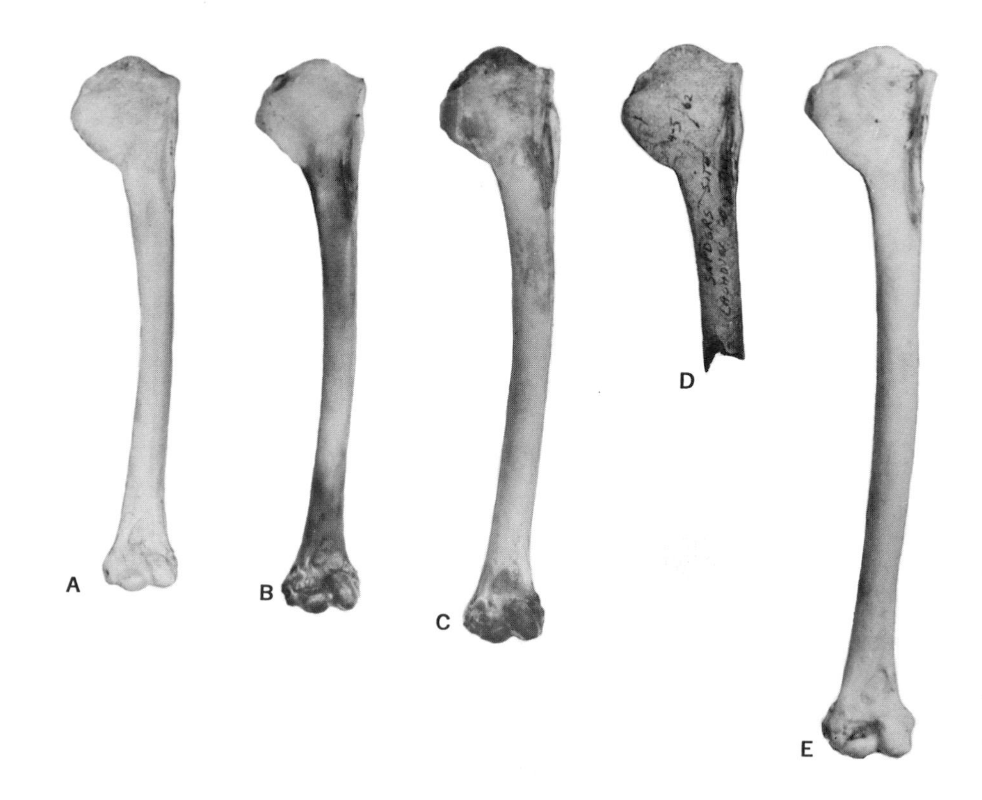

73. *Humeri of:* (*A*) *adult male interior, Horseshoe Lake, Ill.;* (*B*) *adult male maxima, Rochester, Minn.;* (*C*) *adult male maxima, Denver, Col.;* (*D*) *maxima from 2,000-year-old Indian site, Calhoun County, Ill.;* (*E*) *whistling swan.*

74. An *immature of an undescrib race which is slightly smaller th interior, but resembles the gia Canada goose in coloration. Th race's current breeding groun have not yet been determined, b in migration it occurs widely acro the prairies. This example w trapped at the Horseshoe La Game Refuge, Ill., in 1958.*

Superimposed on the influence of the sex hormone in males are the changes in muscle mass directly attributable to the various seasonal stresses. Relative degrees of use and disuse result in varying degrees of hypertrophy and atrophy. During most periods of seasonal stress, the principal muscles of locomotion, as exemplified by the sternal muscles and the muscles of the tibiotarsus, bear a reciprocal relationship to each other. I have considered the wintering goose the basal goose because a normal balance between walking and flying exists at that time. At the end of spring migration, after a minimal flight of 1,100 miles, the weights of these muscles exhibit marked changes in their total mass, the degree of change depending on the age-sex class; the sternal muscles gain 13 to 16 per cent in weight in both sexes of yearlings and in adult females, but only 6 per cent in adult males which, prior to migration, have an 8 to 9 per cent advantage in this respect. (Changes in lipid content of these muscles do not account importantly for these weight changes.) The leg muscles, however, being used less during the migration period, become reduced in size, both absolutely and relatively — markedly in yearlings (9 to 13 per cent), and to a lesser extent in adults (3 to 7 per cent).

During the first few weeks following spring migration, with a resumption of a more normal balance between flying and walking, the sternal and leg muscles return to the approximate mass and relationship of the winter period. The weights of these muscles in females in the early stages of incubation are virtually the same as they were in winter, but their average body weights are 22 per cent less than they were upon arrival on the breeding grounds. Much of this weight reduction is attributable to loss of fat deposits.

The physiological events that take place during the molt, the onset of which is marked by the almost simultaneous loss of the flight feathers of the wing, are probably the most stressful in the life cycle of the goose. It is during this period that a major reorganization of the body tissues occurs; the pectoral muscles, for example, undergo a marked atrophy, the reduction in weight compared with their status at the end of migration being 30 and 36 per cent in males and 25 and 41 per cent in females, for yearlings and adults, respectively. During the winter period, the sternal muscles comprise 20 to 21 per cent of body weight, but at about the midpoint of the growth of the new primary feathers of the wing, these muscles constitute only 14.5 per cent of body weight.

During the flightless period, about 32 days (as measured from the time of loss of wing feathers — actually a few days longer due to the reluctance or inability of a goose to fly when the feathers are becoming loose in their sockets), a goose becomes wholly dependent on its legs for survival. Adults with their young traverse considerable distances each day as they feed, and as the breasts of the goslings become feathered out, the entire family spends longer periods of time on the water. The net result of this increased use of the legs is an extreme hypertrophy of the muscles. At the halfway mark of the renewal of the flight feathers, the muscles of the tibiotarus have increased their weights 41 to 57 per cent, depending on sex and age, as compared with their weights at the end of migration. At the midpoint of the molt the average weight of these muscles of the leg is about 52 per cent of the weight of the sternal muscles as compared with about 25 per cent in winter. By the end of the growth of the primaries, the leg muscles have decreased to their normal size and, simultaneously, the breast muscles have nearly regained their prior mass. These reciprocal changes in weights of the muscles of locomotion during the molt are portrayed in Graph 11.

Metabolic Relationships

The reciprocal changes that have taken place in the weight of the sternal muscles and the weight of the leg muscles by the end of the spring migration, as well as the enlargement of the legs during the flightless period of the molt, clearly represent the influence of relative use and disuse. The marked atrophy of flight muscles during the flightless period, however, has more complex origins. This is indicated by the responses of these muscles in geese held in close confinement. The sternal muscles do not atrophy in caged geese during the greater part of the year despite greatly decreased use; evidently occasional flexing and flapping of the wings in conjunction with preening and holding the wing in place is sufficient to help these muscles retain about normal size and tone. Furthermore, during the regrowth of the primaries, the leg muscles do not enlarge in caged geese as they do in wild geese, as they are put to no greater use during this period. In contrast, the sternal muscles *do* atrophy markedly in caged geese during the molt.

Present knowledge of the nutritional demands of the molt suggest that the breakdown of the tissues of the sternal muscles is an evolutionary development in response to the accelerated need for amino acids for feather growth. This need is accentuated by the fact that it is a *selective* need, the percentage of sulfur amino acids and arginine being much greater in feathers than in muscle tissue. Because neither the sternal muscles nor the wings are of much functional use to geese during the molt, there would be survival value in making selective use of the amino acids degraded from sternal tissues to insure the rapid regrowth of flight

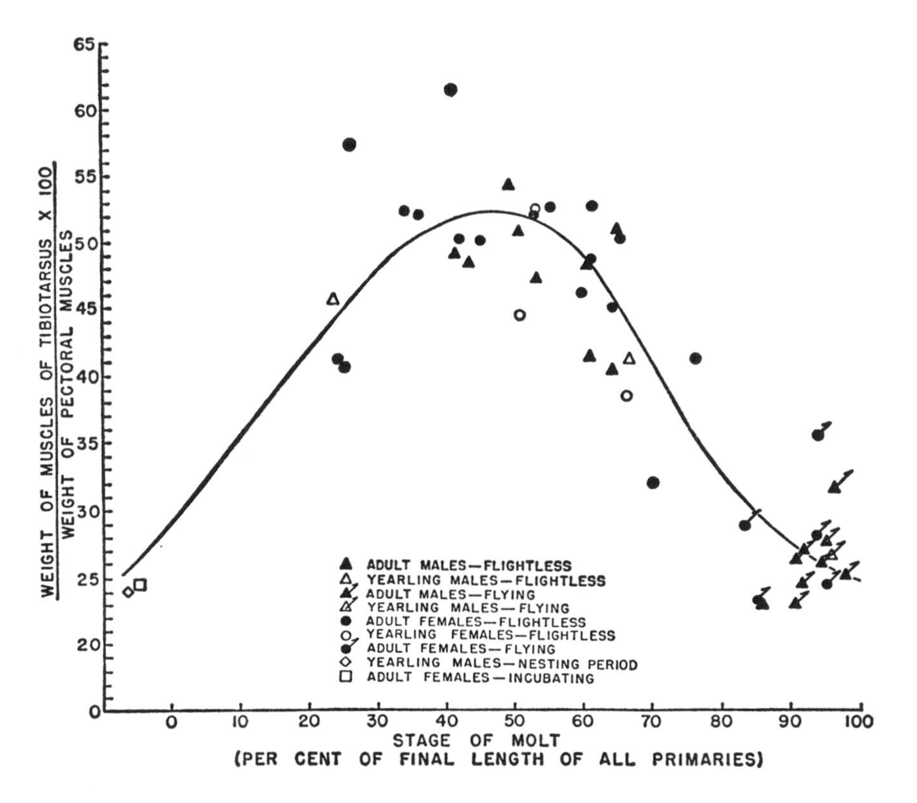

11. *Complementary relationship of weights of muscles of the tibiotarsus to weight of sternal muscles of Canada geese* (B. c. interior) *during the replacement of the primary feathers of the wing.*

feathers, particularly to the 85-per-cent stage of the original total length of all primaries, when flight is again possible. The primary feathers complete their final growth stages more slowly as the sternal muscles are regaining their normal size and as the onset of the molt of the body feathers sets up new demands.

The high demand of the molting goose for sulfur amino acids *appears,* by creating a sulfur deficit per se, to have other ramifications. As a result of the studies of Meister (1951), it is known that birds in molt, particularly waterfowl, undergo extensive destruction of the cortex of the long bones of the legs. The cause and exact mechanism of the development of osteoporosis in molting birds have not been established, but the possibility exists that reabsorption of cortex bone is incidental to the degradation of chondroitin sulfate of the long bones because of a deficit of sulfur in the metabolic pool during the molt. It is perhaps also significant that osteoporosis is more marked in molting adult female geese which have just previously undergone the sulfur-demanding task of egg production. The calcium required for the egg shells is probably not the primary factor in the development of osteoporosis in wild birds, as during the egg laying period medullary bone is temporarily laid down to supply this need. For a more detailed discussion of this possible relationship of sulfur metabolism to osteoporosis, the reader is referred to Hanson (1962*a*).

Mitchell (1962:651–52) has summarized the literature on the basal metabolism of white Leghorn chickens during various seasons of the year. He concluded that the process of egg formation had no appreciable effect on basal metabolism; however, during the autumn molt, the basal metabolism of white Leghorn hens is 43 per cent above the period just prior to the molt and 47 per cent above the period immediately following the molt, as well as for values during the winter period. Clearly, the molt is the most stressful period for the chicken, findings which agree with my conclusions on the significance of the molt in Canada geese (Hanson, 1962*a*).

Weights of Visceral Organs

HEART WEIGHTS — Some of the changes in the weights of the sternal and leg muscles associated with age, sex, sexual maturation, and seasonal stresses in the race *interior* were found to be reflected in the weights of the visceral organs. The weights of the heart muscle of immatures, yearlings, and adults in fall and winter revealed that a significant increase, both absolutely and relatively, occurred in immature males between the time they left their wintering grounds in spring and the time they returned as yearlings the following fall. This increase was attributed, as in the case of the pectoral muscles, to the nitrogen-conserving influence of the male sex hormone, androgen (Hanson, 1958 and 1962*c*). Females, on the other hand, exhibited no increase in heart weight relative to body weight in successive years of life.

SPLEEN WEIGHTS — In the 1962*c* study, I reported that the spleen in Canada geese tends to decrease in size with age. In immatures and yearlings, no differences were found in the relative size of the spleen. In the adult age class, however, the males have a differentially larger spleen, both absolutely and relatively. This may reflect an accommodation to the metabolic requirements of the more highly developed muscular system of this age-sex class. Although the size of the spleen is not directly related to the red cell count of the blood and hemoglobin level, it is of interest to note that the sex differential in the size of the spleen in adult males corresponds with higher hematocrit and hemoglobin values found for this age-sex class (Hanson, 1958). These findings are consonant with the experimental evidence of Domm and Taber (1946) that erythrocyte numbers in the domestic chicken are under hormonal control; hens injected with male hormones have red cell counts similar to males.

The weight of the spleen is, however, highly variable in Canada geese. This may be partly explained by the fact that the spleen of birds can apparently function as a blood reservoir (Harmon et al., 1932).

WEIGHTS OF PROVENTRICULUS AND GIZZARD — A study of the combined weights of the gizzard with the proventriculus attached showed that these organs were, second to the heart, the least variable of those studied (Hanson, 1962*c*). The correlation of the weights of these organs with body weight tended to increase with age, but only in adult females did relationships approach the significant level ($P = 0.1$). The combined weight of these organs, of which the proventriculus accounts for but a small part, is 4.1 to 4.7 per cent of the body weight. For comparison of the relative weights of these and other visceral organs in the race *interior* with other species of birds, the reader is referred to Hanson (1962*c*).

LIVER WEIGHTS — Liver weights were obtained from Canada geese in nearly all stages of the life cycle (Hanson, 1962*a* and 1962*c*). The most meaningful comparison of the differences in liver weights between the various age and sex classes was made from samples collected in autumn which showed that

> the liver averages higher in absolute weight in males than in females, but when compared on the basis of body weight there were no significant differences. The relative weight of the liver in immatures (1.76 per cent of body weight) was markedly higher than in the older age classes, but in both sexes it was relatively greater in adults (1.62 per cent of body weight) than in yearlings (1.44 per cent of body weight). The livers of adults were also found to be more variable in size than in the younger age classes. The best estimates that could be made of the correlation of liver weight to body weight (data pooled by the method of least squares, using the z transformation) were .12 for immatures, .51 for yearlings, and .77 for adults, no difference between the sexes being found. While the significance of the values for immatures is nil, the correlation coefficient increases with age, and in the adults the values are highly significant ($P = .01$) (*from* Hanson, 1962*c*).

At the end of spring migration, the liver weights of adult males averaged 25 to 50 per cent less than the average for the winter period; adult females, on the other hand, had somewhat larger livers — in some individuals they were extremely large, weighing as much as four times the average for the winter period. The largest livers of these adult females in spring were tan in color because of their extremely high fat content. The liver weights of immatures on their return to the breeding grounds showed no important change over their winter weights.

During egg laying, the livers of adult females were markedly decreased in size as compared to their extreme high weights at the end of migration; average liver weights of incubating females averaged 28 per cent less than the average for this age-sex class in winter.

The livers of adults collected during the nonflying stages of the molt of the primary feathers were similar to the weights of livers of wintering geese, but during the flying stage of the molt, the livers of adult geese were 61 and 88 per cent heavier (males and females respectively) than during the flightless stage. In neither stage was the fat content of these livers a significant factor. It is apparent that these two stages of the molt involve basically different metabolic stresses. The biochemical processes involved in the degradation of muscles and the use of some of their constituent amino acids in feather growth, deamination and transamination, are widespread in body tissues and are not confined only to the liver; however, in the metabolism of carbohydrates, the liver plays the dominant role. It is, therefore, not surprising that the liver exhibits a response to the increased energy demands of the goose in the second stage of the molt when flight is resumed. The sudden increase in the size of the liver is also assumed to be a response to the need for energy to drive the formation of protein when the sternal muscles are under rapid reconstruction during the latter stages of the growth of the wing feathers. These immediate energy requirements must be supplied by an increased intake of carbohydrates. This relationship is in contrast to that in males which, not being under the lipogenic stimulation of estrogen, have smaller livers at the end of migration than during the wintering period. Energy for the spring flight of geese is presumed to be supplied largely by their fat reserves, the degradation of which can take place in the muscle tissue and need not directly involve the liver. In adult females, however, the increased size of the liver in spring is an endocrine response geared to egg production.

PANCREAS WEIGHTS — Data obtained on the weights of the pancreas of 119 Canada geese at Horseshoe Lake, Illinois, in autumn showed that this organ constituted between 0.27 and 0.41 per cent of body weight, depending on age and sex. Findings indicated that pancreas weight declined slightly with increased age, both absolutely and relatively, and that relatively, the females had a slightly larger pancreas than males. The pancreas is a highly variable organ in birds, being second only to the spleen in this respect. It was, therefore, not surprising that no significant correlation between pancreas and body weight was found within any age-sex class.

NOTE TO CHAPTER 11

1 The nutritive state of a goose was evaluated from the development and weight of its major muscle groups and the amount of depot fat present.

Behavior 12

THE ADMIRATION and excitement wild geese have evoked in man have stemmed not only from the beauty and spectacular nature of their migratory flights but also from many of their behavior traits, such as the permanency of their pair bonds and their solicitous guardianship of the young. Because of the orderliness of their behavior patterns and the complex hierarchy of their social structure, wild geese have also been attractive subjects for avian behaviorists. It is paradoxical, but fortunate, that the flocks of Canada geese that have been most intensively studied in regard to behavior were *maxima*. Kossack's (1950) study was concerned with nesting and mating habits; the investigations of Balham (1954) and those of Collias and Jahn (1959) were also centered on the behavioral aspects of reproduction but included much on ethology. I have not attempted to summarize or discuss this aspect of behavior as it is one of the central themes of Balham's study which has not yet been formally published. I also believe that if observations on the ethology of Canada geese are to be most meaningfully interpreted, they should be based on comparative data of some of the more diverse races. Balham's study was conducted on the wild flocks at the Delta marshes and at Rennie, Manitoba; Kossack's was on a semicaptive flock at Barrington, Illinois; and the study by Collias and Jahn was carried out on a captive flock at the Horicon Marsh, Wisconsin.

Flock Leading and Flock Leadership

The belief that it is the "wise old gander" that leads the flock has been so long and widely held that it has seemingly become a part of our folklore. It is true, as Balham (1954) has pointed out, and as I have commonly observed at Horseshoe Lake in autumn, that often it is the adult male of a family that initiates local flights. In these instances, preflight synchronization of the family flock is begun by a vigorous up-down tossing of the head by the male and low "gutteral talking." Response by the family to a gander's intentions-signaling varies. In some instances, the other members of the family join the head shaking or may sometimes initiate it; in others, more prolonged signaling by the male, accompanied by short runs with partially raised and spread wings, occurs before the threshold of response for flight is achieved.

Nevertheless, aside from the leadership of the male at such times, careful scrutiny of a family flock on the ground or on the water will reveal that it is often the female that leads the flock and the male (easily identified by his posture and size) that guards it at the rear. It should be borne in mind, however, that merely leading and leadership are two quite different roles insofar as they affect the survival of the family unit. In situations of danger to the family, the male assumes the leadership (Fig. 65).

The leader of large flocks in migration is difficult to ascertain and, in fact, is seen to shift from time to time. Distinct differences in the body size and the wingspread of the family members are easily seen, and it will be noted that the largest birds are not always in the lead position. As to mated pairs, however, it can unequivocally be stated that it is the female that leads the two in the spring migration flight. This is invariably noted by the Cree Indians of the Hudson Bay area when they are forced, because of the difficulty of obtaining other foods at the time of the spring thaw and ice breakup, to hunt Canada geese (*interior*) as they are migrating back to their nesting grounds. In the final leg of their flight to the breeding sites, the mated pairs precede the nonbreeding segments of the population. When these pairs come within gun range, they are shot in quick succession by the Indians. The fallen "lead bird," upon examination, always proved to be the female.

Pair Formation

Many authorities have stated that pair formation in wild geese takes place on the wintering grounds. Until recently, I had held this to be true for *interior*, a conclusion which originated largely from the Indians' reports that these geese are paired on their arrival on the breeding grounds in northern Ontario (Hanson and Smith, 1950). My findings on the breeding grounds in the spring of 1959, however, have indicated that the initial pair-bonds are more likely to be formed on the breeding grounds when the geese are approximately 1 year old. It was found that the testes and the seasonal development of the copulatory organ of young of the previous year (now yearlings) on their return to the breeding grounds were nearly as large as that of fully sexually mature adults (*see* Chapter 8). Undoubtedly, these yearling males are under an endocrine stimulus which is nearly comparable to that of the adults and is sufficient to influence their behavior. My Cree Indian guides in northern Ontario (whom I have found to be astute, reliable observers with insights into biological problems that just skirt technical understanding) have shed some light on this problem. They reported that the geese nested first when 2 years of age, but that they were often in pairs during the second summer of life — or, as my guide, Joseph Chokomoolin of Winisk engagingly phrased it, "just sweethearts the first year."

Balham's (1954) findings in Manitoba and Naylor's (1953) studies in California led them to similar conclusions. Balham observed three 1-year-old female giant Canadas paired with 2-year-old males. He also observed "short term associations between yearlings, or between yearlings and older geese, often accompanied by copulation." He reported two apparent cases of true pair formation occurring in late April. Presumably, copulation in the race *interior* takes place some time during the final stage of migration as I have not observed it at Horseshoe Lake in late winter; 5 weeks later when the pairs reach the breeding grounds, the females are in egg laying condition (Hanson, 1962*a*).

Naylor's (1953) studies of the western Canada goose also led him to conclude that some pairs are formed by Canada geese when yearlings, but that breeding does not take place until they are 2-year-olds.

Collias and Jahn (1959) stated that,

> Successful pair formation in the competition for mates depended both on *specific preferences* and on aggressive *dominance relations* between individuals. One or the other factor might be more important in particular instances. *Preferences* were indicated by persistent tendency of a bird to follow a specific individual. *Dominance* was expressed by defense of the vicinity of the female with special reference to potential or actual sex rivals.

REMATING — It is a common belief that wild geese mate for life and in the event of the death of one of the pair, the surviving member does not remate. Admittedly, owners of captive flocks have reported experiences with individuals of broken pairs that have indicated this; yet it has been apparent from the age and sex composition of heavily shot populations that if members of broken pairs did not remate, these populations could not have maintained their numbers as well as they have (for example, *interior* at Horseshoe Lake, Hanson and Smith, 1950).

Kossack (1950) made the first study of remating of giant Canadas at the Bright Land Farm near Barrington, Illinois. In various experiments involving separation of the mates of 8 pairs, all 16 birds remated. "These mating observations and experiments over a period of two years, resulted in a clear indication that the Canada geese studied will pick new mates if separated from their original mates; the mating usually taking place the next nesting season. In the event that the separation takes place just prior to the nesting season, however, it was found that they will mate the same year."

Balham (1954) concluded from his observations of giant Canadas at Delta: "As far as could be ascertained, no paired goose which suffered the loss of its mate, ever failed to remate."

Sexual Behavior

Collias and Jahn (1959:489) have given an excellent description of the mating behavior of the giant Canada:

> Copulations were often seen to occur between members of a pair before, as well as after, selection of a permanent nesting site by the pair. When they occurred before establishment of a territory, copulations took place in any convenient stretch of open water with a depth of six inches or more. After territorial establishment the location of copulations was invariably in a portion of the pond nearest the nest site. No copulations were observed taking place on land.
>
> Pre-coitional behavior consists of dipping the head deep under the water and then lifting it and at the same time throwing water over the back with the back portions of the head and neck, in a manner similar to the movements of bathing. But, unlike ordinary bathing, the wings are not used; they are kept folded in their normal resting position. This pre-copulatory display, which may be initiated by either the male or female, gradually increases in frequency and intensity over a period of 30 seconds to two minutes or more. The male works closer to the female, and typically grasps the feathers of the back of her neck as he mounts. The female may keep her normal floating position, or she may stretch her neck down at a low angle with the surface of the water partially or completely submerging during the copulation, which lasts only a moment or two.
>
> Post-copulatory behavior is very characteristic. Both birds stretch the neck up high and inclined slightly backward, with the beak tipped upward, and often rotate so that they face one another, breast to breast. Typically the male half raises his wings without unfolding them, in a swan-like pose, and he then gives a snoring vocalization, relatively brief and weak, compared with his ordinary greeting snore to his mate. Meanwhile, the female begins to dip her head vigorously into the water and to lift it again, throw-

ing water over her back in exactly the same fashion as in the pre-coitional display. The male then joins her in this. One or both birds may rear up in the water and flap their wings either before or in between sessions of the head-dipping display. After a few seconds to several minutes of this activity, the birds retire together to a nearby spot along the shore and preen themselves.

Selection of Nest Sites

Klopman observed the initiation of the nesting cycle (*letter*, May 31, 1962):

> As early as three days after arrival [at Dog Lake, Manitoba] one can see nest-site searching by the females. This is true even in the face of snow cover on the islands. Toward the middle of April, where there is some open water around the islands (ice breakup is usually normal May 1) intense fighting between pairs can be seen. There is a gradual increase in the length of time pairs spend on or near the islands from the time of arrival onward. Once the pair spends a considerable time in a particular area of the island, trips back to shore for food stop altogether [Fig. 66].

Collias and Jahn (1959:485) have described the search for nest sites:

> The female usually selects the nest site, leading the male about on exploratory jaunts, getting up on nest islets, and poking about inspecting these heaps of branches, twigs and hay with her beak. But when a desirable nest heap was already occupied by other birds, the male would forge ahead to take the lead in driving the other birds away, and if he was successful the female might then mount the potential nest site and inspect it. The search for a suitable nest site might take one to many days, depending in part on the availability of good nest sites and in part on the dominance status relative to competitors for specific sites. Male AR and his mate got a late start and were rather low in dominance. They found most of the desirable nest sites already occupied. One morning they were seen to visit nine different nest sites in the space of an hour, being evicted from most of these sites by birds that had already laid claim to the sites in question.

Fidelity to nest site appears to be of common occurrence in wild geese. If a nest is successful one year, presumably the same female returns to nest on the identical spot the next year. This trait has been recorded for lesser snow geese, *Chen caerulescens caerulescens* (McEwen, 1958); the greater snow goose, *Chen caerulescens atlantica* (Parmelee and MacDonald, 1960); the pink-footed goose, *Anser fabalis brachyrhynchus* (Scott et al., 1951–1952); and frequently for races of the Canada geese: *maxima* (Kossack, 1950; Balham, 1954); *moffitti*

(Naylor, 1953; Naylor and Hunt, 1954; Hanson and Browning, 1959), and *interior* (*following*).

In examining nests of *interior* in the muskeg country of northern Ontario, I found many nests, both current and of a previous year, that had been built on an earlier nest. In several instances, nest materials and eggshell remnants of two previous nests were found below the one in use. Although it would be a difficult relationship to measure, it might be assumed that the extent of re-use of nest sites by a particular population is roughly related to its mortality rate: the higher the kill and the higher the per cent of old females shot, the more the population is dependent upon females nesting for the first time and the greater the chance that a more random selection of nest sites will occur. This relationship would apply where the habitat is extensive; where nest sites are limited, there would be greater competition for the more suitable sites.

Nest Building and Egg Laying

Collias and Jahn (1959), in their study of the giant Canada goose, have included an excellent description of nest building:

> One female was watched building a nest in a site entirely of her own selection and with nesting materials not provided for her. After rounding out an initial depression in the earth, she gathered nesting material as far as her long neck could reach, stretching out and breaking off small pieces of dead weed stalks, 3 to 6 inches long, with her beak. She then passed each piece of weed stalk from one side to the other across her breast, continuing the movement backwards along one side of her body and dropping the material onto the rim of the nest. After exhausting the supply of nest material within convenient reach she rose and worked her way slowly from the nest to a distance of six feet, breaking off pieces of weeds and dead grass and tossing them to one side and behind herself as she worked along. She would then return to the nest, work in some added material, get up and again work over her previous route, picking up and throwing back nesting material. By this means, material for the nest was moved in toward the nest from the surroundings in successive relays. This nest material was pulled in and then worked into the rim of the nest with the beak while she sat on the nest, until the rim of the nest had reached a height of 5 inches. At intervals she rounded out the nest hollow, turning and shaping it with her breast, and sometimes scraping back with her feet. It took her about four hours to complete the nest (6–10 A.M.), and she then covered over the completed and empty nest with her beak. Within 45 minutes she was found to have laid her first egg in the nest.
>
> Very rarely a male was seen to help the female gather nest material and

build the nest, but even in these rare cases his efforts seemed crude and lacking in intensity.

Nest building continues to some extent throughout the period of incubation. A female, while sitting on her nest, may reach down to the outer and lower border of the nest, pick up nest material in her beak and then deposit it on the rim of the nest close to her body. This procedure, combined with occasional scraping back with the feet used alternately, results in a typical low cone-shape to the outside of the nest, and prevents the nest from becoming flattened down.

As a rule, down is not added until after 3 or 4 eggs have been laid, although different females vary in this respect. Female AA added down after laying only 3 eggs, and on the same day Female W with 4 eggs, and destined to lay a larger clutch than did Female AA, had as yet added no down. The down is plucked from the breast with the beak, and not only down feathers, but some breast contour feathers may be added as well.

On reaching her nest the female almost invariably would stand over the eggs preening her belly and lower breast and apparently working loose some feathers, since a bare area soon appeared from the midline of these areas and increased in lateral extent day by day, during the first part of incubation. The habit of the female of standing over the eggs soon after leaving the water may also help dampen the eggs and improve their hatchability.

Nesting Territories

Both *moffitti* and *maxima* may nest under conditions of crowding but, while the degree of crowding that is accepted probably varies inversely with the size of the local nesting range and the relative availability of nest sites, it will be shown in the following chapter that as crowding increases there is a measurable reduction in productivity. In southeastern Idaho, nesting by *moffitti* on the islands of Blackfoot Reservoir has attained densities of 54 to 66 nests per acre (Jensen and Nelson, 1948); on islands in two lakes in southern British Columbia, Munro (1960) found this race to have nesting densities which varied from 9.4 to 30.7 nests per acre; in northeastern Illinois, a semicaptive flock of giant Canadas attained a nesting density of 11 nests per acre on a small island (Kossack, 1950).

Balham (1954) found that the size of the nesting territory of giant Canada geese on the Delta marshes increased rapidly during the egg-laying period but decreased rapidly after incubation began until finally it included only the nest mound itself. In 1959, I made an extensive study of the nesting of the Hudson Bay Canada goose (*interior*) in the area adjacent to the lower Sutton River which flows into the south coast of

Hudson Bay. Nesting was found to occur in four distinct habitat types but, in all cases, it was widely scattered. Although circumstances permitted observations to be made only during the first third of incubation, it was repeatedly noted that the male would be first observed or flushed at a distance of one-eighth to one-quarter of a mile from the female on the nest. The open character of the nesting areas in northern Ontario may account for these great distances. Balham (1954) concluded that the distances of waiting males from their nests were directly related to the density and height of the vegetation: in the Delta marshes, waiting males were usually less than 100 yards from the nest; at Dog Lake, Manitoba, where nesting is on islands, they sometimes were as much as 300 yards away.

Based on his studies of *maxima* at the Seney National Wildlife Refuge in the Upper Peninsula of Michigan, Johnson (1947) concluded that a desirable nesting density for Canada geese should not exceed two pairs per acre.

RELATION OF SIZE OF NESTING TERRITORIES TO NESTING SUCCESS

The ultimate function of nesting territories in birds is to insure reproductive success. The degree of isolation required by nesting pairs is determined by many factors and varies widely depending on whether or not a species tends to be colonial, but there is ultimately an irreducible limit to crowding beyond which nesting success becomes rapidly reduced. The suppressive effects of crowding on the productivity of Canada goose populations have frequently been reported. Thus, Naylor (1953:90) commented in his findings on *moffitti* in California: "The instinct to defend the territory is never forgotten and when nesting pairs are grouped together in a limited area many fights, quarrels and much loud calling occur among the pairs. This evidently results in many desertions [Fig. 67]."

On one crowded nesting island, Naylor found 31 nests, 11 of which hatched, 4 were destroyed, and 16 were deserted due to crowded conditions or unknown causes. He commented,

> It was on this island that the males stayed close to the females on the nest. The pairs very seldom left the area and then only to feed. There was always a great deal of calling and noise in the vicinity. The pairs were nervous and fighting resulted upon intrusion in the small territories. The writer believes that the overcrowded conditions caused most of the deser-

tions on this particular island. Those pairs that did bring off young often left the nest with the hatching of the first two or three young and deserted the remaining eggs.

A study of *moffitti* nesting on the Tule Lake and Lower Klamath National Wildlife refuges by Miller and Collins (1953:392) yielded parallel findings: "Desertion, the greatest factor causing nest failure in this study, was attributed mainly to intra specific strife on heavily populated islands where competition for nest sites was acute."

Munro (1960:543) reached similar conclusions in his study of *moffitti* in southern British Coulmbia:

> Observed behavior indicates that there is an inverse relationship between territorial competition and the success in hatching young of the individual pairs making up the population. In other words, the breeding success of many pairs is likely to be reduced by the intense territorial competition which occurs when the population is overcrowded. Data relating to hatching success of different time-place groups of clutches . . . also support the contention that territorial antagonism limits success in hatching young of some individual pairs.
>
> Besides the observed behavior there is indirect evidence to support the contention that some females that are ready to lay eggs are unable to occupy sites and build nests in which to lay. Reference is made to numbers of randomly scattered, undamaged eggs found at various points on Goose Island but most commonly on the top. It seems likely that some birds, probably having arrived on the breeding area slightly later than most, were unsuccessful in gaining or maintaining possession of a nesting site and were thereby forced to drop eggs on the ground.

Collias and Jahn (1959:505) have also written in a like vein on the significance of eggs dropped by giant Canadas:

> Perhaps the six dropped eggs found belonged to birds not able to maintain a stable territory, but attempting to lay in some other birds' territory, since all of the six dropped eggs were found within occupied territories in which the resident female laid a full clutch of her own. Furthermore, females were seen on two occasions to be repeatedly attacked by males, just before the female laid an egg in her own nest at the edge of the territory of the attacking male.

These investigators found that (p. 489),

> The importance of the male in territorial defense was made evident by cases in which there was no male to defend the nest site. Female W/Y lost her mate by death late in her incubation period, and although she had been

sitting very steadily up to the time the male was lost, her eggs failed to hatch. The reason was that with loss of her guardian she became subject to the domination and disturbance of other pairs as well as of unpaired males, who drove her repeatedly from her eggs, resulting in death of the embryos, presumably from chilling.

Collias and Jahn (p. 502) attributed the success of experienced birds in part to their relatively high dominance status:

This . . . brings out the importance of factors concerned with failure to lay eggs, at least in nests. Over half the birds assumed to be capable of breeding failed to make nests in which they laid and incubated eggs; one-fifth of the birds failed even to pair up effectively. The nine pairs that failed to lay eggs in nests were involved in unusually frequent territorial clashes, and most of them were unable to maintain stable territories for any length of time. Four of the 5 pairs (and perhaps all 5) that lost their clutches did so because of disturbance to the female from other geese, related to a lack of effective male defense. It would seem that over half the loss in breeding potential could be ascribed to factors having to do with territory, i.e., to lack of effective territorial establishment or defense. . . . The possible alternative is that many of the females that were sufficiently motivated to defend territories and to copulate, for some reason were not sufficiently in breeding condition to lay eggs. These two explanations are not completely antithetical.

BROODING

"During the first few days after leaving the nest, the parents would usually lead their brood back to the original nest to be brooded at night [Collias and Jahn, 1959:495]." Similar observations were made by Kossack (1950). To some extent, brooding on the nest site may reflect crowding and confinement as Balham (1954) found no indication of this habit on the Delta marshes.

In general, Collias and Jahn (1959:496) found that "During the first week, the goslings were brooded quite often during the day; after the first week it was evident from general observation that they were brooded less frequently, although no data were assembled on this point. All of the females brooded their downy goslings when it rained, and at night."

MOVING TERRITORIES

Once the young have hatched, the family becomes what Jenkins (1944:35) has described as "a family supraorganism, since it performs the activities of a larger, more complex individual, through coordination of its components. This results in the dominance of the family, which is of survival value to its members in that they can feed first and rest in the center of the aggregation and are not pecked or chased." The gander now defends the area which immediately surrounds the family as it moves and feeds, and as Jenkins has pointed out, the defended area thus becomes a moving territory (Fig. 68).

The Integrity of Goose Families

The foregoing discussions have assumed that the assemblage of young that accompanies the adults to the wintering grounds is the same as that hatched by these adults the previous spring. My studies of *interior* over a 20-year period, both at Horseshoe Lake and on their breeding grounds in northern Ontario, have yielded no data whatsoever to indicate any important deviation from this working hypothesis; rather, there is a great deal of evidence to support the above assumption. At Horseshoe Lake, the average size of flocks in flight was determined for a number of years. Our counts were limited to flocks of nine or less because this was the largest group I had previously trapped, color-banded, and subsequently observed on the ground as a family unit. From this study, the following conclusions were reached (Hanson and Smith, 1950:172):

> 1. That the average family-flock size in late summer or early autumn may furnish a rough index of the age ratio within a large population . . . from this ratio the success of nesting the previous spring may be inferred.
> 2. That the average family-flock size in middle or late autumn, when compared with similar data gathered the same year before the opening of the hunting season, is indicative of the degree to which family groups have been broken through shooting.

The finding that average flock size could be related to age ratios and productivity has formed the basis of a continent-wide study of the productivity of goose populations (Lynch and Singleton, 1964:114–37).[1]

If in northern Ontario the original broods at hatching did not remain as distinct entities but became re-sorted as random assemblages of young, surely this would have been indicated by observations on the breeding grounds, or by observation of family flocks and by the ratio of adult females to immatures caught in traps in autumn. Also, since 1959, Harry G. Lumsden and I have taken hundreds of exceptionally clear photographs of family goose flocks during low-level plane flights over the breeding grounds. These photographs, taken for production studies, have provided ratios of adults to young that were within expected limits.

However, the alternate hypothesis has been advanced, presumably by Boyd (*in* Scott et al., 1955:87), that family flocks of geese observed in autumn are essentially a random assortment of young resulting from loose brood bonds on the breeding grounds. He reasoned (from England) that because there were reports of brood aggregations of Canada geese (*moffitti*) at a few areas in our western states, similar conditions must necessarily exist in the vast muskeg breeding grounds of northern Ontario. In the opinion of Henry A. Hansen (*personal communication*), who has made extensive studies of Canada geese (*moffitti*) in the West, the parents of young in large aggregations rotate their activities between caring for the goslings and feeding. If this opinion is confirmed, a normal ratio of attendant parents to young would actually prevail in such local flocks. However, it should be pointed out that massed aggregations of broods of *moffitti* are by no means the rule in the West. For example, Kebbe (1955:162) made counts of broods in eastern Oregon throughout the 1954 breeding season and obtained the following average sizes: 13 Class I broods, 5.2; 61 Class II broods, 4.1; 33 Class III broods, 3.8. It can only be concluded from these data that Kebbe was observing families that had continued to remain a distinct entity. He has also provided some excellent information on the rate of brood attrition from various mortality factors.

What are the social forces that preserve the integrity of the brood? Isolation of the breeding pairs generally permits family bonds to be established before contact with other broods would permit realignment of the young. This may sometimes take place when the goslings are very young — a viewpoint with which Balham (1954) is in agreement. Although the hostility expressed by the parents in driving off stray goslings from other broods is also a factor in maintaining brood integrity, I am convinced that the *strength of the bonds between the young themselves* is equally important in maintaining the continuity of the original brood. On Akimiski Island in 1958, 75 to 100 goslings from 10 to 21 days old were held in pens for 1 to 2 days prior to banding and release. When these young were fed bundles of grass, the competition for food made the different broods (which could be distinguished by their size, development, and cohesive behavior) very intolerant of each other, manifested by their frequent and aggressive pecking.

Balham (1954) made a careful study of flocking behavior of giant Canada geese on the Delta marshes, stating that "The characteristic behavior during the first stage was strictly confined to the family level.

Avoidance of other families was the rule, with no following or intermingling. Few contacts were seen at this stage."

Balham noted that while groups of broods gradually exhibited a considerable degree of social cohesiveness in their response to their changing environment, unity of the individual family was maintained. "Regardless of how or when the different families were integrated into the flocks, each family remained an entity within the general structure. Further, the subordinate non-family birds were not tolerated within the limits of the family flock area, and arranged themselves around the periphery."

The Collias and Jahn (p. 496) observations support my own as to the importance of gosling bonds in maintaining brood integrity:

> Parent birds, as well as goslings a week or more old, would attack strange goslings that ventured near, especially if there was a wide discrepancy in size from goslings in the family, permitting ready recognition of strangers as such. Adoptions of strange goslings were most likely to occur when both own and strange goslings were less than one week old, and of about the same size. These younger goslings have apparently not yet become thoroughly acquainted and attached to a specific parent bird, nor their parents to them. Adoption is also facilitated by dominance of the foster parents. When a gosling accidentally gets into another brood and is allowed to remain with that brood, its parents may be thwarted in their endeavors to get it back, by the aggressive attacks of the other parents defending the vicinity of their own brood. Usually, however, the original parents succeed in calling their own goslings out of another brood with which they might have intermixed, and goslings possibly recognize the voices of their own parents. Whether or not the gosling stays with and is permanently adopted into the other brood depends in part on the degree of attachment of the gosling to its own parents, and in part on the degree of attachment or indifference of the parents to their goslings, as well as on other circumstances, such as the degree of tolerance or intolerance of the foster family towards the newcomer. Adoptions were rare in the case of goslings more than one week of age.
>
> As the goslings grew older, their aggressiveness greatly increased. When less than a week old they would hiss at a human captor and, when over a week of age, battles between goslings were not infrequent. *Strange goslings from other families were particularly prone to be attacked*, but fights sometimes involved members of the same brood. It was not long before the goslings seemed to share, to some extent, in the dominance status of their parents, and it was remarkable to observe a downy gosling only about two weeks old forge well ahead of its father to drive an adult gander away from the food hopper [Italics by present author].

STATUS OF YEARLING GEESE IN SUMMER

Jenkins (1944) and Balham (1954) both found that young (yearlings) of the previous nesting season may rejoin their respective parents some time after the latter have hatched their new broods. Extensive aerial observations in the James-Hudson bays area by Harry G. Lumsden and me suggest that some of the yearlings in the population of *interior* behave similarly, but I believe that most yearlings in wild populations of *maxima* and *interior* do not. As pointed out in Chapter 4, "Migration," there is considerable evidence to substantiate the belief that a large portion of the yearling giant Canadas summer on the Arctic barrens; the yearling population of *interior* in the Hudson Bay region tends to segregate on large lakes either within the treeline or on adjacent areas of Arctic tundra. Many of the yearling and nonbreeding adults that migrate to Akimiski Island in James Bay in the spring spend much of their flightless period in large flocks on open water at the east end of this island.

INTER-FAMILY DOMINANCE

As the goslings mature, the bonds between the young themselves and between the young and their parents strengthen and, as Balham (1954) has suggested, this increased cohesion of the family probably has considerable survival value. Accompanying this increased intra-family attachment is a seeming awareness of family size on the part of the adults. It was noted when making a behavior study of *interior* at Horseshoe Lake, Illinois, that the outcome of conflicts between ganders of two different families *did* have a pattern which, although unrelated to the size and apparent strength of the ganders involved, was predictable on the basis of the size of the opposing families. In most instances, the gander of the larger family routed the gander of the smaller family. Quite obviously, the peck order or "social standing" of the gander was related to the size of his family (Hanson, 1953). These observations were subsequently confirmed by Boyd (1953) in his study of white-fronted geese (*Anser a. albifrons*) in England. From Balham's (1954) observations of families of giant Canada geese during the period of growth of the goslings, he concluded that the age (size) of the goslings tended to influence the outcome of threat-encounters and actual conflicts, and also that the gander initiating the encounter was usually the victor. In rebuttal, it should be noted that my conclusions were based on observations in autumn and winter when the goslings of different broods no longer varied appreciably in size.

The hierarchy within aggregations of geese is more extended and complex than one existing solely between families competing for food and space. It was repeatedly noted in studies of *interior* at Horseshoe Lake in the 1940's (when the use of walk-in traps required many hours of observations in a blind) that when conflict situations arose between two or more individuals, the following descending peck-order was the general rule: families were superior to mated pairs without families; mated pairs to single adults and yearlings; and yearlings to immatures. The lone immature separated from its family was clearly at the bottom of the peck order and therefore at a disadvantage when food was either limited in amount or when it was concentrated, as in the case when corn was fed in winter (Hanson, 1953). This order of social relationship has been confirmed by Balham (1954) in his study of giant Canada geese at Delta, Manitoba, and by Boyd (1953) in his investigation of the white-fronted goose in England.

General Behavior and Disposition

The placid disposition of the giant Canada goose sets it apart from all other races and, as Hinde and Tinbergen (1960) and Mayr (1960) have pointed out, behavioral traits constitute completely acceptable taxonomic characters.

The tendency of giant Canada geese to remain apart from other races of Canada geese on the wintering grounds has been noted. Mershon (1925) writes,

> Even after the big flight came from the North and the White-fronted or Laughing Goose, the Snow Goose, and Canada [*interior*] and his smaller brother, the Hutchins, had come in thousands and thousands and were literally covering the fields at feeding time, *these big geese held aloof and did not mix with the others* [Italics by present author].

Veteran hunters in southern Manitoba also reported to Robert McWhorter in 1962 that the giant Canada goose usually keeps apart from other races of Canada geese when both are present on the same feeding grounds. Harvey K. Nelson and I observed an instance of this kind at Lac Qui Parle, Minnesota. A flock of several thousand migrant *interior* was seen basking in the sun along the lake shore, while on an island a short distance away a flock of seven giant Canadas was feeding by themselves.

The tameness of the giant Canada goose as compared with other races often impressed hunters: "On this point [near the Arkansas River in Kansas] were about a hundred very large geese, of a lighter color than the rest, and they did not seem so timid" ("Widgeon," 1922:23).

When these geese winter at refuges in towns (such as the Inglewood Sanctuary in East Calgary, or the city parks at Rochester, Minnesota, and Denver, Colorado), they soon become quite tame and many will feed at the feet of an observer and, at times, from the hand.

The resident population of Canada geese at the Seney National Wildlife Refuge, Michigan (which the present study has proved to be *maxima*), differs markedly in behavior from migrant flocks (*interior*) which stop there (Johnson, 1947:23):

> These [immatures] undoubtedly were refuge-bred geese as exhibited by their fearlessness of the refuge personnel and the readiness with which they fed with the pinioned birds. Approximately 2,000 "wild" geese [*interior*] in several small flocks also used the refuge water areas during the autumn flight, but were easily distinguished from the "refuge" birds by their extreme wariness and disposition to take wing when approached.

It is this inherent tameness that has permitted the giant Canada goose to be so readily domesticated. Accounts of early explorers mention tame Canada geese around Indian villages, and the first settlers often kept them about their homesteads and killed them in autumn for food. The tameness of this race also made it the ideal bird to use for live decoys; virtually all captive flocks in the north-central states originating from decoy flocks have proved to be *maxima*. If this race of geese had been native to Europe, it would undoubtedly have been the progenitor of a domestic breed.

Vocalizations

The following analysis of vocalizations is from Collias and Jahn (1959:498–99).

> Some ten or more different vocalizations were observed in Canada Geese [*maxima*]. These sounds are listed below, together with the situation that seemed to provoke each. This list is not intended to be exhaustive.
>
> 1] *Hiss.* Directed against other geese, ducks, and especially against humans near the nest [Fig. 69]. Apparently signifies threat and alarm at *short distances.* For example, one female on her eggs repeatedly hissed at the observer when he approached within a few feet, and then switched to honking when he moved off, reverting to hissing each time the observer returned close to her.
>
> 2] *Honking.* The honking of the male is loud and resonant, and each honk seems to be relatively prolonged, as compared to the short, staccato

honking or yipping of the female. It is probable that spectographic analysis would reveal characteristic differences in the frequency components in the voices of the two sexes. Geese characteristically honk in at least five different situations: 1) in territorial advertisement and warning to intruders, 2) as a *long-distance* call or answer to the mate when separated by more than just a few feet or yards, 3) as part of the greeting ceremony when mates come together after having been separated, 4) as an alarm call when a man or dog, for example, approaches a wild flock, or when a gosling is threatened or seized by a man and gives its distress calls, and 5) in flight or when about to take flight. The honking seems to be of the same essential character in all of these situations, although it is possible that significant nuances of this vocalization under different circumstances have escaped the observer. There seem to be two common elements in the various stimulating situations, first, some element of *alarm*, and secondly, functioning of the call over some *distance*. In the case where geese honk in flight one may reasonably assume that fear of separation from companions provides the occasion for honking, just as a flightless male may honk down his free-flying mate when she takes wing. The alarm and the distance functions of honking may help to explain the loudness of this call compared to others, and its strident and high-pitched character.

3] *Short-distance call of the mate.* This call is a low, short, rather soft grunt, given repeatedly, about once a second: *kum! kum! kum!*, etc. Sometimes it is double-noted, sounding like *wah kum!*, the first note being low and brief and the second, higher and abrupt. A free-flying male was observed to give the short-distance low call on trying to get his wing-pinioned mate to fly off with him, but when he actually began to run into the wind and took off, he switched to the usual loud honking. Another example may be cited. A female left her nest and eggs to drink, and then she swam toward the food site, while the male trailed up to 50 feet behind her repeatedly calling *kum! kum!* in an apparent effort to induce her to return to her nest. He continued to lag, and soon the mates were separated by a considerable distance, whereupon he suddenly began to honk loudly, and swam rapidly to join the female as she came near the feeding area. Both sexes give the short distance call.

4] *Short distance call to the goslings.* This is a rapid series of short, low, soft grunts, quite similar to the preceding call, but often faster and not so loud. Possibly it is the same call. Either sex may give it, but it is most often heard from the female.

5] *Special greeting call for the female.* This rather loud, prolonged, snoring vocalization is peculiar to the male who directs it only to his mate. It is heard when a pair has just formed, and whenever the male and female come together after a period of separation, whether or not any aggression against other geese was involved. Possibly it functions to reinforce the pairing bond.

6] *Post-copulation call.* Immediately following copulation the male gives a brief, light snore, while assuming the characteristic posture that has been described in preceding pages.

7] *Scream of pain.* An abrupt, high, rather short scream, heard when one gander was taken by surprise and received an unexpected bite from another.

8] *Distress call of adult.* A loud *oh!-oo, oh!-oo,* etc., heard when one bird was seized, held and bitten by a dominant bird. It was sometimes heard when a bird was separated from its mate by a fence.

9] *Distress call of gosling.* A loud peeping, typically given when a gosling was lost, or was removed from its family by the observer.

10] *Contentment notes of gosling.* A light, rapid series of soft notes: *wheeoo, wheeoo,* etc., given for example, when the gosling was returned to its family.

NOTE TO CHAPTER 12

1 I have not pursued this type of analysis in studying the productivity of *interior* at Horseshoe Lake because I believe the information obtained is less reliable than that obtained from trapping. Furthermore, as pointed out in Chapter 13, "Productivity and Regulation of Populations," production data secured from photographic surveys on the breeding grounds combined with information on the age and sex composition of a population in winter now provide much more meaningful assessments of productivity.

Productivity and Regulation of Populations 13

RARELY, or for only comparatively short periods of time under special circumstances, do animal populations achieve theoretical rates of reproductive efficiency. What is of concern in management is whether or not normal reproductive rates are being achieved, how much they vary from year to year, and the cause of these variations. The chief objective of this chapter is to describe the reproductive capacities of *maxima* so that population fluctuations can be better understood and hunting regulations can be adjusted to anticipated population levels. Efforts to understand the dynamics of Canada goose populations have heretofore been somewhat handicapped by the lack of statistics on the age and sex composition of unshot or lightly hunted populations. I believe the data for the Rochester flock help to fill this hiatus as these geese receive only light hunting prior to their arrival at Rochester and practically none thereafter. Information available on the productivity of *maxima* populations proved to be considerable; some of these data had been published, but most of it was supplied by game biologists in various governmental agencies. These statistics together with published data on populations of *moffitti,* which I have summarized for comparative purposes, have provided a new understanding of the regulation of Canada goose populations.

Age at Sexual Maturity

The age at which Canada geese first nest is important in evaluating the reproductive potential of a population. Observations of captive flocks of Canada geese by game breeders have indicated repeatedly that most females do not breed until 3 years of age or until they enter their fourth summer of life. Confinement per se, or its corollary — the prevention of normal migration or prenesting flight activity — may in some way be intimately linked with an inadequate stimulus of the endocrine system. Comparative data on wild and captive flocks of *moffitti* studied by Craighead and Stockstad (1964) have supported the concept that confinement reduced the rate at which Canada geese became sexually mature. It has long been my belief, however, that the Mississippi Valley Flyway population of *interior* could not have sustained the degree of harvest as well as it has in certain past years unless a significant number of the 2-year-old females in the population nested. (Lynch and Singleton [1964: 122] have recently reached similar conclusions in regard to blue goose populations.) Some concrete evidence for this belief was obtained in northern Ontario in 1959 (*see* Chapter 8, "Characters of Age, Sex, and Sexual Maturity").

Brakhage (1962) has accumulated excellent data on the free-flying flock of giant Canada geese nesting at the Trimble Wildlife Area, Missouri, and in the surrounding countryside. Of 22, 2-year-old males and 25, 2-year-old females present in the flock in 1962, 11 of the females (44 per cent) nested and 14 of the males (64 per cent) were mated to nesting females. It is apparent from these data that the 2-year-old age class *does* make substantial contributions to the annual productivity of Canada goose flocks. Recently Craighead and Stockstad (1964) have shown that one-third of the 2-year-old geese (*moffitti*) in the Flathead Valley of Montana nested. I suspect, however, that the percentage of 2-year-old geese that breed in any given year in areas where nesting habitat is of limited extent tends to be inversely related to the per cent of available breeding territories in those areas that are occupied by older breeding pairs.

Wood (1964) has incorrectly cited Balham (1950). Balham found that only a small percentage of 2-year-old (*not* 1-year-old) geese nested, but as he had data on only a few individuals which had been held in captivity during the winter, he regarded his findings as inconclusive. Wood (1964) found that only about 5 per cent of the semidomesticated Canada geese (*maxima*) obtained from the Michigan Conservation Department Game Farm nested at 2 years of age but, as these geese were introduced to their pens the same breeding season they were under study and were held in fairly close confinement under varying degrees of deliberate crowding, it is doubtful that his study has much validity for wild populations.

It was noted at Horseshoe Lake, Illinois, that after the heavy kill of immatures (*interior*) that took place in 1943, the population decreased markedly in 1945. The year-class of 1943 should have made a substantial contribution to the flyway population in 1945 (Hanson and Smith, 1950). This two-year delayed effect of a heavy kill of immatures was also noted in several subsequent years (Hanson, *unpublished*). Similar relationships between the relative strength of various year classes and rates of population increase are evident in the census data for the Rochester flock of *maxima* and will be discussed below.

Clutch Size

Clutch size varies with age; it has been generally recognized by game breeders that Canada geese lay fewer eggs the first time they nest than they do in subsequent years. Dutcher (1885) cites a game breeder on Long Island who claimed that four eggs are laid the first year of breeding, five the second, and six or seven thereafter. Miner (1923) stated that "a young goose will lay four eggs the first year of laying and usually five the second." In my studies of *interior* in northern Ontario (*see* Chapter 8), the only 2-year-old nesting female collected had a clutch of three eggs. The nests of seven other older females collected all had clutches of at least five eggs.

Brakhage's (1962) study of the Trimble flock of *maxima* in Missouri has provided the first substantial data on the relation of clutch size to the age of the female in Canada geese: the 2-year-old females had 11 successful nests with an average of 4.7 eggs per clutch; 3-year-old females had 6 successful nests with 5.5 eggs per clutch; and females 4 years old and older had 26 successful nests with 5.8 eggs per clutch.

No significant difference was found between the data for the average clutch size for *maxima* and *moffitti* (Table 22). The average clutch size for *interior* may eventually be shown to be somewhat higher than for the above races, but the data presently available are inadequate to substantiate this relationship (Table 22).

Nesting Success

A successful nest is one which has produced one or more young. The populations of *maxima* that have been studied by various workers in respect to nesting success fall into two principal categories: (1) flocks of all wild birds (in Manitoba, Saskatchewan, South Dakota, Wisconsin, and Missouri); and (2) flocks composed of all semicaptives or partly so (in Wisconsin, Illinois, and Ohio). It was not deemed important in this summary to distinguish between the two categories in computing average clutch size, but it was desirable to do so in considering other aspects of the nesting cycle.

For all populations of *maxima*, the average rate of nesting success was 58.6 per cent, or 82.4 per cent of the average rate for *moffitti* (71.1 per cent, Table 22). The average rate for only free-flying, wild populations of *maxima* was 53.3 per cent, or 75.0 per cent of the rate for the populations of *moffitti*, all of which were wild. Captive stocks of *maxima* in Wisconsin had an average nesting success of 71.9 per cent, a rate 22.7 per cent higher than that for the wild populations of *maxima* and nearly identical to the average nesting success for wild populations of *moffitti*.

Hatching Success

The rate of hatching success for all eggs laid by a Canada goose population, to a large degree, reflects nesting success, but for reasons discussed under "Behavioral and genetic factors," the differential between these two evaluations of reproductive success is sometimes considerable. Nevertheless, the average hatching success for *maxima*, 55.8 per cent, and *moffitti*, 75.5 per cent, does not differ greatly from the average rates of their respective nesting success. No difference between the average hatching success of successful nests of *maxima* and *moffitti* is indicated by the data available (Table 22), but the data for *maxima* is inadequate and therefore a meaningful comparison is not possible.

FACTORS AFFECTING NESTING SUCCESS

The factors that determine the outcome of the nesting cycle may be divided into two generalized categories: (1) external, or environmental, and (2) internal, or behavioral and genetic. In many instances of nest failure it is manifestly impossible to make sharp distinctions between the two. Frequently, an apparent primary cause of nest failure proves, upon closer study, to be a secondary cause. Despite this basic difficulty and errors inherent in trying to separate the causes of nesting failures into these divisions, an attempt must be made to do so if an insight is to be gained into the population dynamics of Canada geese, particularly an understanding of the underlying causes for the marked differences in the nesting success of *maxima* and *moffitti*.

ENVIRONMENTAL FACTORS — Manifold ecological factors exert suppressive pressures on Canada goose reproduction (Table 23). In general,

22. Clutch size, nesting success and hatching success for populations of *Branta canadensis maxima*, *B. c. moffitti*, and *B. c. interior*[1]

Subspecies and province or state	Area	Year
B. C. MAXIMA		
Manitoba	Dog Lake	1954–55
Saskatchewan	Along South Saskatchewan River	1962
South Dakota	Waubay N.W.R.	1943–62
Wisconsin	Horicon Marsh[6]	1952
Wisconsin	Horicon Marsh[6]	1950–57
Wisconsin	Crex Meadows[6]	1958–63
Illinois	Bright Land Farm, Barrington	1945–46
Missouri	Trimble Wildlife Area	1962
Ohio	Mosquito Creek Area[6]	1959–63
Total or average		
B. C. MOFFITTI		
British Columbia	Upper Columbia Valley and Southern Okanagan Valley	1949–53
Washington	Columbia River	1953–56
Oregon	Malheur N.W.R.	1938
California	Honey Lake	1939–40
	Honey Lake	1951
	Susan River and Honey Lake	1952
	Tule Lake and Lower Klamath N.W.R.	1952
Washington	Lower Columbia River	1950
Idaho	Blackfoot Reservoir	1946–48
Idaho	Grays Lake	1949–51
Montana	Flathead Valley	1953–54
Utah	Bear River N.W.R.	1937
	Bear River N.W.R.	1946–48
	Ogden Bay	1947–48
Total or average		
B. C. INTERIOR		
Ontario	Sutton River Area	1959

1. Data in parentheses are derived from data of "authority." 2. Weighted average. 3. Tub nests; excluded from averages. 4. Ground nests. 5. These data excluded from column totals and averages on clutch size and eggs hatched.

these factors tend to be density-independent rather than density-dependent, but again no absolute division between the two can be applied. But it should also be realized that the environment operates on the genetic background of an animal; to paraphrase a human adage, the environ-

Number of nests studied	Number eggs produced	Average clutch size	Per cent of nests successful	Per cent of eggs hatched		Authority
				All nests	Successful nests	
104[7]	(474)	5.1	46.0	51.4	96.0	Klopman (1958)
						James L. Nelson
38[5]	—	—	42.0	—	—	(*letter*, January 3, 1963)
384[5]	—	—	58.3	—	—	Robert R. Johnson
						(*letter*, October 16, 1963)
23[5,8]	133[5]	5.2[5]	69.6[5]	53.4[5]	(85.5)	Collias and Jahn (1959)
111	545	4.9	72.1	57.4	—	Richard A. Hunt (*unpublished data*)
74	328	4.4	71.6	69.2	—	Richard A. Hunt (*unpublished data*)
140	729	5.0	57.0	55.6	—	Kossack (1950)
72	336	5.4	75.5[3]	61.4[3]	(81.5)[3]	Brakhage (1962)
			63.2[4]	48.0	(78.7)	
204	1,083	5.2	—	53.3	—	Kenneth E. Allen
,127	3,495	5.0(5.0)[2]	58.6	55.8	86.7	(*Progress Report*, 1963)
378	1,907	5.0	—	79.9	—	Munro (1960)
,032	(5,582)	5.4	70.9	(64.8)	92.0	Hanson and Browning (1959)
			63.0	—	—	Sooter (1938)
418	1,812[9]	5.1	56.9	—	93.1	Dow (1943)
360	1,904	5.5	68.3	(59.2)	82.6	Naylor (1953)
						Naylor and Hunt (1954)
115	432	5.2	72.2	(59.9)	82.9	
201	810	5.1	78.6	(68.4)	87.0	Miller and Collins (1953)
282	(1,579)	5.6	—	—	—	Hansen and Oliver (1951)
296	1,553	5.2	—	55.9	—	Jensen and Nelson (1948)
361	(1,893)	5.2	80.0	(67.2)	86.0	Steel, Dalke and Bizeau (1957)
358	1,912	5.3	78.6	55.6	89.5	Geis (1956)
84	410	4.8	—	80.7	—	Williams and Marshall (1937)
459	2,336	5.1	—	85.8	—	Jensen and Nelson (1948)
29	152	5.2	—	77.6	—	Jensen and Nelson (1948)
,373	22,282	5.2(5.1)[2]	71.1	75.5	87.6	
9	50	5.5(5.6)	—	—	—	This report

6. Captive flock. 7. Clutch size data for 93 nests. 8. Included in figure immediately below and hence excluded from total. 9. Based on data from 355 nests.

mental pressures upon an animal may sometimes not be as important as how the animal responds to them.

Flooding is an important agent of nest destruction in some areas; at Dog Lake, Manitoba, 23 per cent of the nests of giant Canada geese were

23. Factors limiting nesting success in populations of *Branta canadensis maxima* and *B. c. moffitti*

Subspecies and area	Environmental factors limiting nesting success (per cent)							
	Flooding		Predation		Miscellaneous or unspecified		Total (per cent)	
	Eggs	Nests	Eggs	Nests	Eggs	Nests	Eggs	Nests
B. C. MAXIMA								
Manitoba								
Dog Lake		23.0		13.0		7.0		43.0
Wisconsin								
Horicon Marsh					(7.5)[1]	8.7	7.5	8.7
Illinois								
Bright Land Farm, Barrington			10.4		25.1		35.5	
Missouri								
Trimble Wildlife Area			8.6		3.0		11.6	
Ohio								
Mosquito Creek								
Average		23.0	8.0	13.0	5.3	7.9	18.2	25.9
B. C. MOFFITTI								
British Columbia								
Upper Columbia Valley and Southern Okanagan Valley	1.2		3.2		2.0		6.4	
Washington								
Columbia River		3.0		13.0				16.0
California								
Honey Lake	8.9			19.9		5.0		33.8
Honey Lake		1.6		4.3		1.6		7.5
Susan River and Honey Lake				16.5				16.5
Tule Lake and Lower Klamath N.W.R.		5.0		2.5		2.5		10.0
Idaho								
Blackfoot Reservoir					32.5[2]		32.5	
Grays Lake		4.2				1.4		5.6
Montana								
Flathead Valley		2.4		22.2				24.6
Utah								
Bear River Marshes	2.2		4.4		2.4		9.0	
Bear River Marshes					9.9[2]		9.9	
Ogden Bay					22.4[2]		22.4	
Average	5.6	3.2	3.8	13.1	13.8	2.6	16.0	16.

1. Figures in parentheses are derived from data of authority. 2. Apparently also includes infertile eggs and eggs with dead embryos; hence data not included in total

lost during the 2-year-period of study by Klopman (1958). In a similar period at Honey Lake, California, Dow (1943) found that flooding destroyed 8.9 per cent of the nests of *moffitti* under observation (Table 23).

Behavioral and genetic factors limiting nesting success (per cent)						
Desertion		Displaced or dropped eggs	In-fertile eggs	Eggs with dead embryos	Total egg failure[3]	Authority
Eggs	Nests					
	11.0		4.0		4.0	Klopman (1958)
(18.8)	(21.7)	4.5	7.5	8.3	15.8	Collias and Jahn (1959)
			3.0	12.9	15.9	Kossack (1950)
9.2	19.4		5.1	15.8	20.9	Brakhage (1962)
	26.5					Kenneth E. Allen (*Progress Report,* 1963)
14.0	19.7	4.5	5.2	12.3	14.2	
8.4		1.8	1.8	2.1	3.9	Munro (1960)
	11.0		2.0	6.0	8.0	Hansen and Browning (1959)
	6.9		2.3	4.6	6.9	Dow (1943)
	23.9		2.0		2.0	Naylor (1953)
	11.3		2.1		2.1	Naylor and Hunt (1954)
	11.4		1.9	9.4	11.3	Miller and Collins (1953)
11.7						Jensen and Nelson (1948)
	15.0		7.0	4.0	11.0	Steel, Dalke and Bizeau (1957)
	13.9			6.1[4]	6.1	Geis (1956)
3.9			6.3		6.3	Williams and Marshall (1937)
4.2						Jensen and Nelson (1948)
						Jensen and Nelson (1948)
7.1	13.3	1.8	3.2	5.2	6.4	

column. 3. Includes only last 2 columns; data based on findings for *successful* nests. 4. Includes both infertile eggs and dead embryos.

Moderately low temperatures usually do not reduce nesting success, but where a population (*moffitti*) nests in a climatically marginal area, late spring storms may be a critical factor. "At Blackfoot Reservoir the heaviest rainfall comes later in the breeding season [than at the Bear

River refuge] during May and produces maximum effects on breeding geese because the storms come during the critical incubation period. Furthermore, they can and usually do come as hail and snow and deserted nests which have been matted down by excessive moisture are often observed (Jensen and Nelson, 1948)." In the area cited above, nesting failures were due directly to weather condition and not desertion.

I suspect that drought, unless severe, does not seriously affect the productivity of Canada goose populations; however, Russell R. Hoffman has reported (*letter*, December 31, 1963) that in 1961 drought had an adverse effect on the nesting population at the Bowdoin National Wildlife Refuge.

In presettlement times, fire may have been a depressive factor on the nesting of Canada geese on the Great Plains but it would be of little consequence now.

Many species of birds and mammals cause the destruction or abandonment of the nests of Canada geese but the species chiefly responsible for nest losses vary widely, as would be expected, in different localities. Heavy losses in an area from one species of predator may seemingly suggest the desirability of control measures, but the classical studies of Errington (1946 and 1963) serve to remind us of the futility of such programs. Indeed, we need look no further than the data in Tables 18 and 19, particularly in the latter which shows the comparative constancy of nesting success and especially the constancy of hatching success in the various populations of *maxima* and *moffitti*, to realize that the different species of similar kinds of predators complement each other in their predation from area to area and that, basically, there is an intercompensatory relationship between the various categories of predators as well as between all agents of nest destruction.

In the words of Errington (1946:235),

> we may see that a great deal of predation is without truly depressive influence. In the sense that victims of one agency simply miss becoming victims of another, many types of loss — including loss from predation — are at least partly intercompensatory in net population effect.
>
> . . . predation looks ineffective as a limiting factor to the extent that intraspecific *self-limiting mechanisms* [italics by present author] basically determine the population levels maintained by the prey.
>
> Even in equations depicting predator-prey interactions in lower vertebrates, loss types may substitute naturally for each other instead of pyramiding, and compensatory reproduction should not be ignored when a resilient

instead of a rigid fecundity is indicated [*see* section "Regulatory Mechanisms" in this chapter].

In the far west and on the Great Plains, the coyote (*Canis latrans*) may be the chief mammalian predator; in the east and northeast portion of the range of *maxima*, the red fox (*Vulpes fulva*) supplants the coyote in importance. Crows, ravens, and magpies may similarly replace each other as chief avian predators.

Unfortunately, little information is available on the predators responsible for the destruction of nests of giant Canada geese on the Great Plains. This race, because of its superior size, can be presumed to have an advantage over *moffitti* in coping with the larger mammalian predators (*see* below). At Dog Lake, Manitoba, the red fox, when stranded on the forested islands after the melting of the ice, was found to be the principal agent of nest destruction (Klopman, 1958).

In his study of *moffitti* at Honey Lake, California, Dow (1943) found that "coyotes [*Canis latrans*] were responsible for more nest destruction than were all other predators combined." Naylor (1953), in a subsequent study at Honey Lake, cited the coyote and the striped skunk (*Mephitis mephitis*) as the chief mammalian predators of the nests of the western Canada goose; black-billed magpie (*Pica pica*), crow (*Corvus brachyrhynchos*), ring-billed gull (*Larus delawarensis*), and California gull (*Larus californicus*) were the principal avian predators. Geis (1956) determined that over 90 per cent of nest destruction in the Flathead Valley, Montana, was due to crows and ravens (*Corvus corax*). Munro (1960) assigned all nest losses on islands in two lakes in southern British Columbia to crows and black-billed magpies, and Hanson and Browning (1959) found that black-billed magpies were responsible for the great majority of nests destroyed on islands in the Columbia River in Washington.

A population composed of individuals of superior size and strength can be expected to have a statistically greater expectation of survival against a predatory enemy than would a race of smaller-sized individuals. The survival advantages of *maxima* in the face of attack by a predatory mammal can be inferred from the following episodes. Mr. Kenneth Brossmann of Oakes, North Dakota, related to me that when he shot and wounded the 18¾-pound goose shown in Figure 8, his terrier rushed in and seized it by the neck. Despite the fact that one of the wings of the goose had been broken, it was able to lift the terrier 6 feet into the air in an attempt to shake it loose and escape. A somewhat similar account was given to me by Robert R. Johnson, manager of the Waubay refuge. A hunter in his area had also wounded and downed a large gander. In this instance, the hunter's cocker spaniel rushed in to retrieve it, only to be knocked senseless from a well-directed blow on the head from the bent wing of the wounded goose. (*See* Figure 58 for wing "spur" which serves as an effective defensive weapon.)

To summarize, the data available do not permit any firm conclusion as to the relative importance of various environmental factors in controlling the breeding success of populations of *maxima* and *moffitti*. Average figures indicate no significant difference exists between these subspecies, but the average for *maxima* is heavily weighted by losses due to flooding at Dog Lake, Manitoba. Allowing for the variations in ecology of the study areas, and variations in the methods of reporting the data, no marked over-all difference in the rate of nest predation on either *maxima* or *moffitti* populations is evident.

BEHAVIORAL AND GENETIC FACTORS — Again, no strict distinction between these two categories is possible; for example, a population containing a relatively high percentage of 2-year-old females nesting for the first time may have a higher rate of nest loss due to desertion than a population composed largely of old, experienced, and fully sexually mature females. Geis (1956:415) has pointed out that nests with large clutches of eggs were usually better hidden and were less frequently deserted or destroyed than nests with four eggs or less. She (correctly, I believe) attributed this differential in nesting success to differences in the ages of the females. Thus, the rate of nest desertion would reflect a difference in individual behavior due to age rather than a behavior pattern of genetic origin characteristic of the entire population. Distinctive behavior patterns during nesting in the population as a whole could, of course, be of genetic origin. The lack of data on these aspects for wild populations of *maxima* further emphasizes the need for interpreting figures in Table 23 with caution. If it can be assumed that data for captive and semicaptive populations are at least indicative of the behavioral and genetic background of nesting success in wild populations, then it would appear that these factors contribute to the low rates of nesting success achieved by some *maxima* populations as compared with those of *moffitti*. Unfortunately the causes of nest failures in the wild in Saskatchewan and South Dakota are not known.

Nest and egg losses from desertion are higher for *maxima* than for *moffitti*. Populations of both races are intolerant of crowding, but it is reasonable to suspect that *moffitti* has become genetically more adapted to it than *maxima*. Wood (1964) has shown from experimental studies of penned geese (*maxima*) that crowding during the breeding season is a factor that can consistently inhibit reproduction. In relation to its over-all breeding range, which is largely in semidesert or mountainous terrain, the nesting habitat of *moffitti* has always been relatively restricted. The giant Canada, on the other hand, once had an enormous over-all range but one which only in the better sectors might have supported nesting concentrations comparable to *moffitti*. In the early days of hunting in North Dakota, giant Canada geese were usually observed in small family parties rather than in large flocks, a fact which in itself would suggest that scattered nesting was the rule on the north-central prairies.

The incidence of both infertile eggs and embryonic mortality is markedly higher in *maxima* populations (Dog Lake, Manitoba, population excepted) than in those of *moffitti*. While I doubt that the rates of infertility and of dead embryos are generally as high in most wild or in completely unrestricted populations of *maxima* as in captive or semi-captive populations, the data *do* suggest these aspects of nesting may account in part for the differentially lower rate of productivity in *maxima* than in *moffitti*. Indirect support for this conclusion is contained in Banko's (1960:130) findings on the trumpeter swan (*Cygnus cygnus buccinator*):

> In 1949, for instance, 30 of 61 swan eggs laid in 12 nests failed to hatch — a loss of 49 percent. In 1951, the loss was 34 percent in the 13 nests checked, when 25 eggs remained in the nest from a total of 73 incubated. Examination of 178 eggs in 36 nests in 1955 revealed that 36 percent failed to hatch for one reason or another.

Banko points out that similar egg losses have been reported for the trumpeter swan in the Kenai Peninsula, Alaska, and for the mute swan (*Cygnus olor*) in Europe. While he is unable to assign the relative roles played by infertility and mortality of the embryo in the trumpeter swan egg, he concluded that, "Loss of productivity due to egg failure is a major factor in the present low production of cygnets on the [Red Rock Lakes] Refuge."

These data on egg losses from infertility and embryo mortality in Canada geese, swans, and rates known for ducks suggest clines related to size, longevity, and rates of population turnover, and to some extent, the evolutionary age of the genera, species, and races concerned. Selection pressures for high egg success has doubtless also been greater for small forms which necessarily suffer nest destruction from a greater variety of predators. According to Banko, predation of trumpeter swan nests is not important to over-all hatching success.

Brood Size

Information on brood size is substantial (Table 24) but it has the shortcoming that not all of it is strictly comparable. In the instances where I have derived figures from related data, the brood size is the theoretical average size at hatching; this was calculated by dividing the number of goslings hatched by the number of successful nests. However, most of the brood observations of *maxima* made in the wild relate to goslings at least one-half to three-quarters grown. Allowing for the unevenness in the quality of these data, it would appear that the average brood size of *successful* pairs is about the same for both races — or possibly slightly higher in *maxima*. This conclusion is somewhat in conflict with data on the higher rate of infertility and embryonic mortality in *maxima* which would be expected to result in lower average brood size. The lower rates of productivity of *maxima* populations as compared with those of *moffitti* are particularly evident when the number of goslings produced by various flocks is related to the total number of pairs that nested. That a significant difference between the productivity of the two races does exist is also indicated by the constancy of the differences between the data for the two races (Table 24).

Compensatory Mechanism

The low rate of productivity of *maxima* populations must be compensated by higher survival rates and perhaps greater potential longevity. Records of captive Canada geese living for 30, 40, and even as long as 80 years have been cited by Hanson and Smith (1950:188). It is realized that most such records of extreme longevity, not too uncommon for captives, pertain to old decoy stocks of giant Canada geese. In the wild, however, few if any individuals would attain the potential life span of the species.

At this point it may be appropriate to briefly reply to the question commonly asked by the layman, "How long do Canada geese live in the wild?" The answer must necessarily be based on findings on *interior*, the first waterfowl population to be studied in respect to population turnover and longevity. Band recoveries from the bandings at the Jack Miner Bird Sanctuary, Kingsville, Ontario, from 1925–49 and from the bandings at Horseshoe Lake from 1940–45 were used in the analyses. The composite life-table method of analysis which I originally developed in 1945 for this study (Hanson, pp. 172–88 *in* Hanson and Smith, 1950) and elucidated in great detail in footnotes to the tables in the 1950 report, was subsequently used by Bellrose and Chase (1950), Hickey

(1952; 1959:342), Nelson and Hansen (1959), and numerous other workers in similar studies of other waterfowl populations.

Average survival, and to a lesser extent the survival of a few individuals for a decade or more, now depends upon the severity of hunting losses. A population reduced through heavy losses from hunting may compensate by having a higher rate of productivity, but the net effect is a high rate of turnover in the population. In contrast, a stable, lightly hunted population can be expected to exhibit low reproductive rates because of the saturated status of the nesting range, but the individuals in it to have a high average survival, with an appreciable portion of the population living to comparatively old age.

In the early 1940's, when the Horseshoe Lake population was declining under tremendous annual losses from hunting, the average immature did not survive long enough to breed (2 years of age); it was computed that the mortality rates for immatures the first year after banding was approximately 74 per cent. For older geese, the mortality rates varied from approximately 56 to 66 per cent. Prior to this period of heavy hunting losses at Horseshoe Lake, life tables based on band recoveries from the Miner Sanctuary, 1925–32, showed that the average mortality rate after banding for geese of *all ages* was about 55 per cent. These data also indicated that about 95 per cent of a year-class in the Mississippi Valley Flyway population was dead 7 to 8 years after banding. It is probable that even if this population was lightly hunted the turnover period would not be appreciably lengthened. While a few individuals in the wild may live 20 or more years, the number that attain this age is only a fraction of 1 per cent of a year-class. A band recovery record of such an individual was recently received. This goose, a female, was banded at Horseshoe Lake, Illinois, on March 1, 1942 when an immature; on March 17, 1964, it was found dead near Rantoul, Illinois.

Rates of Population Increase

The Rochester flock has been censused since 1952 (Table 16). The statistics indicate very irregular rates of increase, possibly in part as a result of varying degrees of accuracy in the censuses and from failure, in some years, to find all of the wintering population.[1] Yet, the population increased in 13 years from a relative handful of birds to 6,000, an increase which was probably largely the result of self-contained productivity rather than ingress of other breeding flocks from increasingly larger areas of the breeding grounds of this race in Manitoba. The most simple analysis of these census data was to average the per cent of population change from year to year, with the assumption that this method would

24. Average brood size and number of goslings produced per nesting pair by populations of *Branta canadensis maxima* and *B. c. moffitti*

Subspecies and province or state	Area
B. C. MAXIMA	
Manitoba	Dog Lake
	Alf Hole Refuge
	Cedar Lake
	Lake Winnipegosis
	Waterhen Lake
	Lake St. Martin
	Lake Manitoba
Subtotal or average	
Montana	Russell N.W. Range
South Dakota	Waubay N.W. Refuge
Minnesota	Agassiz N.W. Refuge
Wisconsin	State-wide (see Fig. 39)
	Horicon Marsh[5]
	Horicon Marsh[5]
	Horicon Marsh
	Dodge County
	Barkhausen Sanctuary and Bay Beach Wildlife Area
	Crex Meadows Wildlife Area[5]
	Crex Meadows Wildlife Area
Illinois	Bright Land Farm, Barrington
Missouri	Trimble Wildlife Area
Ohio	Mosquito Creek Area[5]
Subtotal or average	
Total or average	
B. C. MOFFITTI	
British Columbia	Upper Columbia River Valley and Southern Okanagan Valley
Washington	Columbia River
California	Honey Lake
	Susan River and Honey Lake
	Tule Lake and Lower Klamath N.W.R.
Idaho	Blackfoot Reservoir
	Grays Lake
Montana	Flathead Valley
	Upper Snake River
Utah	Bear River Marshes
	Bear River Marshes
	Ogden Bay
Total or average	

1. Or productive pairs. 2. Or average number of goslings produced per successful pair. 3. Figures in parentheses are derived from data of authority.

Year	Number of broods[1]	Number of goslings	Average brood size[2]	Goslings produced per nest or nesting pair	Authority
1954–55	(48)[3]	244	(5.1)	(2.3)	Klopman (1958)
1963	12	38	3.2	—	H. P. Law
1947	36	134	3.7	—	Arthur S. Hawkins (*from river basin surveys*)
1947	11	47	4.3	—	Arthur S. Hawkins " " " "
1947	4	11	2.5	—	Arthur S. Hawkins " " " "
1947	3	14	4.7	—	Arthur S. Hawkins " " " "
1947	26	115	4.4	—	Arthur S. Hawkins " " " "
	140	603	4.3 [4.0][6]	(2.3)	
1963	30	(129)	4.3	—	Kenneth F. MacDonald
1943–62	224	1,054	4.7	2.7	Robert R. Johnson (*unpublished refuge data*)
1957–63	208	1,013	4.6 [4.9][6]	—	Herbert Dill (*unpublished refuge data*)
1948–63	27	(148)	5.5	—	Richard A. Hunt (*unpublished data*)
1952	16[4]	71[4]	(4.4)[4]	3.1	Collias and Jahn (1959)
1950–57	77	293	3.8	—	Richard A. Hunt (*unpublished data*)
1952–63	85	376	4.4	—	Richard A. Hunt (*unpublished data*)
1952–57	48	197	4.1	—	Richard A. Hunt (*unpublished data*)
1954–61	190	734	3.9	—	Richard A. Hunt (*unpublished data*)
1958–63	53	227	4.3	—	Richard A. Hunt (*unpublished data*)
1956–63	45	206	4.6	—	Richard A. Hunt (*unpublished data*)
1945–46	—	—	4.1	—	Kossack (1950)
1962	40 (from tub nests)[4]	159[4]	(4.0)[4]	2.6[4]	George Brakhage (1962)
	12 (from ground nests)	37	(3.0)	1.6	George Brakhage (1962)
1959–63	—	—	—	2.9	Kenneth E. Allen (*unpublished data*)
	1,139	5,017	4.2 [4.4][6]	2.6	
	534	3,129	4.2 [4.6][6]	2.5	
1949–53	—	—	—	4.0	Munro (1960)
1953–56	732	3,617	(4.9)	(3.5)	Hanson and Browning (1959)
1951	(246)	1,127	(4.6)	(3.1)	Naylor (1953)
1952	96	(403)	(4.2)	(3.1)	Naylor and Hunt (1954)
1952	158	705	4.5	3.5	Miller and Collins (1953)
1946–48	(296)	868	(2.9)	—	Jensen and Nelson (1948)
1949–51	285	1,257	4.4	3.5	Steel, Dalhke and Bizeau (1957)
1953–54	416	1,130	2.7	(2.7)	Geis (1956)
1947	—	—	4.0	—	Craighead and Craighead (1949)
1937	—	—	—	(4.0)	Williams and Marshall (1937)
1946–48	1,262	5,474	(4.3)	—	Jensen and Nelson (1948)
1947–48	(29)	118	4.1	—	Jensen and Nelson (1948)
	3,520	14,699	4.1 [4.2][6]	3.4	

4. These data excluded from totals and averages. 5. Captive flock. 6. The figures in brackets are weighted averages.

25. Age and sex composition of the population of giant Canada geese wintering at Rochester, Minnesota, as indicated by trap catches in January, 1962 and 1963, and February, 1964

Age-sex class	Number trapped 1962	1963	1964
Immature males	24	77	48
Immature females	16	72	46
Sub-total and per cent	40	149	94
Yearling males	14	12	8
Yearling females	15	20	9
Sub-total and per cent	29	32	17
Adult males	20	62	32
Adult females	13	48	22
Sub-total and per cent	33	110	64
Total	102	291	165

tend to cancel the errors of observation and provide a figure reasonably close to the actual average annual rate of increase. This calculation yielded an average rate of increase from January to January of 33.1 per cent (Table 16), a figure which suggests that mortality in the Rochester flock of *maxima* has been exceptionally low. A low mortality rate is plausible from several standpoints: (1) losses to the population from hunting in recent years apparently have been low; (2) the population has gradually reoccupied breeding range long vacated; (3) a relative lack of competition for nest sites; and (4) low, over-all population densities, a factor which would reduce the potential for the transmission of diseases and parasites.

The Breeding Component, Recruitment, and Mortality

If trap samples of a wintering population are of sufficient size as to be statistically reliable and if the trapping methods used do not differentially select for any one age or sex class, the data on the age and sex structure of the flock which these samples provide are useful in calculating the breeding success achieved the previous spring, the number of young that are likely to be produced in the following spring, and the size of the flock that can be reasonably expected to return to the wintering grounds a year later (Hanson and Smith, 1950:192). Thus if figures on production are lacking, projected estimates of the population from one winter to the following autumn should be based on what has proved to be the average performance of the flock.

The trap samples examined at Rochester, Minnesota, for age and sex composition were taken with cannon-net traps. There has been some concern, however, that catches made with these traps do not accurately

Per cent			Immatures per adult female			Ratios Males per 100 females		
1962	1963	1964	1962	1963	1964	1962	1963	1964
23.5	26.5	29.1						
15.7	24.7	27.9						
39.2	51.2	57.0	3.1	3.1	4.3	150 : 100	107 : 100	104 : 100
13.7	4.1	4.8						
14.7	6.9	5.5						
28.4	11.0	10.3				93 : 100	60 : 100	89 : 100
19.6	21.3	19.4						
12.8	16.5	13.3						
32.4	37.8	32.7				154 : 100	129 : 100	145 : 100
100.0	100.0	100.0						

reflect the age-sex composition of the population. Therefore, the data presented in Table 25 may not constitute wholly reliable samples of this flock, particularly because of the small size of the total annual catches. In my opinion, they do represent adequately the structure of the population. The percentage of immatures — a commonly used yardstick of productivity — can, however, be a misleading statistic as it is influenced by annual variations of the sex ratio of the sexually-mature segment of the adult population and by annual variations in the percentage of 1½-year-old geese. The representation of this latter age group is, in turn, a reflection of the number of young produced 1½ years previous and the mortality this age class suffered in the interval, but particularly in the first year of life.

A projection of the maximum size of the breeding component in the Rochester flock that could return to the breeding grounds in a given spring was calculated as follows: (per cent yearling females + per cent adult females) × 2 × number of wintering geese. In making this estimate it was realized that not *all* of the 2-year-old geese breed at this age, that some of the older adults may not breed, and that an unknown percentage will fail to nest successfully.

Thus (steps condensed) from Tables 15 and 24:

From January 1962 data: 55.0 per cent × 4,000 = 2,200 in spring of 1962

From January 1963 data: 46.8 per cent × 5,300 = 2,480 in spring of 1963

From February 1964 data: 37.6 per cent × 6,000 = 2,256 in spring of 1964.

The number of breeding geese (excess adult males disregarded) in the flock the ensuing winters was calculated as follows: per cent of adult females in the flock × 2 × the January census figures.

Thus (steps condensed) from Tables 15 and 24:

From January 1962 data: 25.6 per cent × 4,000 = 1,024 in spring of 1961

From January 1963 data: 33.0 per cent × 5,300 = 1,749 in spring of 1962

From February 1964 data: 26.6 per cent × 6,000 = 1,596 in spring of 1963.

Only the calculations for 1962 and 1963 can be compared. The projected maximum size of the breeding component in the spring of 1962 was calculated to be 2,200 geese; sex and age data the following winter indicated there were 1,749 of these geese left in the flock. This suggests a mortality rate of 20.5 per cent (figures actually relate to the combined fortunes of the yearling and adult female components of the flock). The adult segment of stable populations of wild geese would usually experience heavier mortality rates, but in view of the sharp gain made by the Rochester flock in 1962, the above estimate may not be greatly in error.

The projected size of the breeding component for the spring of 1963 was 2,480; the calculation made above from the 1964 trap data indicated that the potential breeding population had been reduced to 1,596 geese by midwinter, 1963–64, an indicated mortality rate of 29.3 per cent. Regardless of whether or not this mortality rate is based on fortuitously biased trap samples, it is probably a fairly close estimate of the mortality rate of yearling and adult females in goose populations during a period of stability or moderate increase.

The trap samples indicated that the number of immatures per adult female in the Rochester flock from 1962–64 was 3.1, 3.1, and 4.3 (Table 25). The average production rates indicated by the data for 1961 and 1962 would seem to be realistic estimates of actual production. While I was at first skeptical of the high rate of average production indicated for the spring of 1963 by the February 1964 trap data, it may not be greatly amiss. The recent studies which Harry G. Lumsden and I made on the breeding grounds of *interior* have shown that brood size may be exceptionally high in some years, exceeding average values by as much as 30 per cent. At Dog Lake, Manitoba, which lies near the center of the breeding range of the Rochester flock, Klopman (1958) recorded 129 young produced from 44 nests in 1954 and 115 young produced from 60 nests in 1955, or average production rates of 2.9 and 1.9 per nesting female, respectively. The average number of young produced per *successful* nest in those same years was 4.8 and 5.5, respectively (data in Table 24 are averages of both years).

Data for the flock at the Waubay National Wildlife Refuge in northeastern South Dakota (Table 26) are uniquely valuable because of the span of years they cover and for the understanding they provide into the population structure of an indigenous stock of giant Canada geese residing in the heart of its range on the Great Plains. This flock has fluctuated considerably in numbers from year to year, but basically it has been stable. For example, from 1945–49, the average number of pairs on the refuge was 16; during the last 5 years of the period covered by the data, 1958–62, the average number of pairs on the refuge was 17. The data do, however, have several weaknesses: the number of yearlings that may have flown to the Arctic or nearby areas outside the refuge to molt prior to the spring counts is not known; also, in some years, the number of yearlings and adults counted on the refuge in spring at the start of the nesting season exceeds the total number of geese, including goslings of the year, reported to be at the refuge during the previous summer. Thus, egress or ingress and incomplete counts were likely some of the factors responsible for the discrepancies, but these are presumed to be partly balanced out and compensated by using the overall totals and averages in the calculations below.

At the Waubay refuge, 40.5 per cent of the total cumulative population for all years attempted to nest, but only 23.9 per cent were successful in rearing broods — a nesting efficiency of only 59.0 per cent from the standpoint of the number of pairs *participating* in the breeding season (Table 26). The *potential* size of the total breeding population was probably somewhat larger than 40.5 per cent of the entire population, as it is reasonable to assume that in all years a certain, but unknown, percentage of the pairs failed to even *attempt* to nest due to competition for nesting niches and territorial conflicts.

The total of all annual recruitments to the Waubay population from 1945–62 was equal to 56.7 per cent of the total cumulative populations present in those years at the start of the nesting season. The average over-all mortality rate sustained between midsummer and the beginning of the following breeding season was 37.5 per cent. These data are of interest because the Waubay flock was essentially stable within the range of its variation after the first 2 years of nesting (1943–44).

The total known kill of Canada geese within the vicinity of the Waubay refuge, 1950–62, was 686 birds, a number equivalent to 92.1 per cent of the total recruitment to the flock in the corresponding period.

26. Size of breeding components, gosling production, and mortality rates in populations of *Branta canadensis maxima* and *B. c. moffitti*

Subspecies and locality	Years	Total yearlings and adults in spring	Nesting adults: Number	Nesting adults: Per cent
B. c. maxima				
Waubay N.W.R.,				
South Dakota	1945	92	24	26.1
	1946	70	28	40.0
	1947	130	30	23.1
	1948	138	20	14.5
	1949	186	58	31.2
	1950	164	136	82.9
	1951	74	52	70.3
	1952	74	30	40.5
	1953	60	30	50.0
	1954	65	48	73.8
	1955	166	40	24.1
	1956	60	40	66.7
	1957	50	30	60.0
	1958	102	30	29.4
	1959	137	30	21.9
	1960	89	34	38.2
	1961	115	40	34.8
	1962	50	40	80.0
Total or average (weighted)		1,822	740	40.6
B. c. moffitti				
Flathead Valley,				
Montana	1953	800	400	50.0
	1954	1,075	432	40.2
Total or average		1,875	832	44.4
Gray's Lake,				
Idaho	1950	1,000	600	60.0
	1951	800	500	62.5
Total or average		1,800	1,100	61.1

Undoubtedly, migrants (*maxima* as well as *interior*) accounted for part of the kill, but it is nevertheless true that hunting pressure on this population has been severe (Harvey K. Nelson, *personal communication*). A causal relationship is generally evident between the size of the kill in the Waubay area for the various years and the status of the flock on the refuge in the spring 1 and 2 years later (Table 26). In 1950, a kill equivalent of 56.2 per cent of the population was reported; a year later, the yearling and adult population present in the spring had declined by 54.9 per cent. The population was static in 1951, but declined markedly again in 1952 as the result of 2 consecutive years of heavy kill. However, the severity of heavy hunting losses in 1950 had been partly compensated by

Productive adults		Number of goslings produced	Recruitment to population (per cent)	Population change from midsummer to following spring	Population trend (total yearlings and adults $\frac{[\text{year 2} - \text{year 1}]}{\text{year 1}} \times 100$)	Geese shot in vicinity of refuge	
Number	Per cent					Number shot	Kill expressed as a per cent of refuge population
20	21.7	58	63.0				
18	25.7	43	61.4	−53.3	− 23.9		
24	18.5	53	40.8	+ 6.2	+ 46.2		
16	11.6	43	31.2	−24.6	+ 6.2		
38	20.4	91	48.9	+ 2.8	+ 34.8		56.2
46	28.0	110	67.1	−40.8	− 11.8	154	31.3
28	37.8	70	94.6	−74.5	− 54.9	45	36.7
22	29.7	46	62.2	−48.6	0.0	44	9.7
16	26.7	33	55.0	−50.0	− 23.3	9	20.8
24	36.9	60	92.3	−30.1	+ 8.3	26	49.4
30	18.1	79	47.6	+32.8	+155.4	121	24.0
30	50.0	90	150.0	−75.5	− 63.9	36	59.3
8	16.0	9	18.0	−66.7	− 16.7	35	32.6
18	17.6	33	32.4	+96.6	+104.0	44	36.9
24	17.5	61	44.5	+ 1.5	+ 34.3	73	34.8
24	27.0	49	55.1	−55.1	− 35.0	48	12.1
30	26.1	67	58.3	−16.7	+ 29.2	22	33.0
20	40.0	38	76.0	−72.5	− 56.5	29	
							35.2
436	23.9	1,033	56.7	−37.5		686	
328	41.0	634	79.3				
259	24.1	496	46.1	−25.0	+ 34.3		
			60.3				
587	31.4	1,130					
450	45.0	990	99.0				
430	53.8	1,010	126.3	−59.8	− 20.0		
880	48.9	2,000	111.1				

the high rate of recruitment in 1951. Light hunting losses in 1953 and moderate losses from hunting in 1954, combined with a high recruitment rate in 1954, made possible the sharp gain in the population in 1955. The degree of the recovery indicated for 1955 is, however, much exaggerated due to error in the censuses or an ingress of 2-year-old geese which, when yearlings the previous summer, were probably molting off the refuge and hence not counted, as the data indicate a 32.8 per cent gain from midsummer of 1954 to the spring of 1955 instead of a decrease that must have occurred. A heavy kill in 1955 was followed by a downward trend in the population in 1956 and 1957, but again cushioned by a high rate of recruitment in 1956 and by more moderate losses from hunting. No direct links between the number of geese bagged in the area of the Waubay refuge in 1957 and population trends are apparent.

Conclusions derived from the Waubay data are necessarily tentative, both because of inherent shortcomings in the records and as a result of evaluating field data second hand. It would also be misleading to relate fluctuations of this population entirely to the effect of local hunting as Canada geese nesting in the surrounding area, as well as migrants from farther north, undoubtedly were included in the recorded kill. Although it would appear that mortality from local hunting reduced the Waubay population in some years, it is likely that losses from hunting partly supplanted losses from natural agencies that would otherwise have occurred. Furthermore, it will be shown below that reductions of this population were compensated by higher rates of productivity.

It would be reasonable to assume that a high percentage of immatures (50 to 60 per cent) in a goose flock in autumn would necessarily indicate a thriving and secure population. But more than likely it may point to a population whose age structure is out of balance as a result of heavy losses from hunting or from one or more preceding years of extremely low productivity, or a combination of both. If hunting were nonselective of any particular age class, it would not alter the age structure of a population, but studies at Horseshoe Lake, Illinois, and elsewhere, have repeatedly shown that hunting results in a disproportionately high kill of immatures. Thus, in years when total hunting losses are high, the immature age class is severely decimated (Hanson and Smith, 1950). As a result, in years following heavy kills, the adults of breeding age and their young would each comprise a proportionately higher percentage of the total population — by virtue of the relative absence of yearlings — than would be the case if hunting were nonselective.

Data for two populations of *moffitti* (Table 26) nicely illustrate the relationships of mortality occurring between the summer of one year and the spring of the next and the relative importance of annual recruitment to the age structure of goose populations. It should be first noted, however, that both of the populations of *moffitti* contained a higher percentage of geese that attempted nesting than did the Waubay flock of *maxima*, and that the proportion of these (*moffitti*) that nested successfully was higher. For these reasons, recruitment rates in both populations of *moffitti* were also considerably higher than for the Waubay flock. But between late summer of the first year and the spring of the second year of the respective studies, the histories of these two populations of western Canada geese contrast sharply: the Flathead Valley geese suffered an over-all calculated mortality of only 25.0 per cent (this figure is probably lower than the actual) and were therefore able to increase by 34.3 per cent; the Grays Lake population, on the other hand, either as a result of a poor breeding season the previous spring or heavy losses suffered from hunting the previous fall, had a high recruitment rate (probably not as high as indicated due to failure to account for all yearlings and excess adult males derived from this population which may have flown to other areas for molting). Despite a high level of recruitment, the Grays Lake flock declined by 20.0 per cent from the first spring of the study to the second spring due to a mortality rate of nearly 60 per cent occurring between the summer of the first year and the spring of the second year. It is reasonable to believe that the high proportion of immatures in the Grays Lake population in 1953 set the stage for the high mortality rate (59.8 per cent). It is also noteworthy that despite the varying mortality rates of these two flocks, their total populations at the end of the second breeding season were not greatly different from their numbers at the end of the first: Flathead Valley flock, 1,571 from 1,434, a 9.6 per cent increase; Grays Lake flock, 1,810 from 1,990, a decline of 9.0 per cent.

Ultimate Factors of Productivity

The multiplicity of factors — internal and external — that affect the productivity of goose populations from year to year have been discussed in some detail. Most of these parameters of productivity are, however, merely tributary to the two ultimate factors that determine the over-all reproductive performance of a population: (1) the per cent of the potentially sexually mature birds in the population that are productive (because of the unbalanced sex ratio that usually prevails in favor of the males, only the adult female component of a population need be considered); and (2) the average number of young per pair reared to a flying or an independent stage. (This value is probably satisfactorily provided by data obtained on broods at the latest age at which the young can still be visually separated from the parents.) The literature, as has been shown (Table 24), is replete with data on brood size, although some of it is unsatisfactory because of the early age of the goslings when brood size observations were made or, in some cases, because I derived figures for average brood size from the authorities' base data. The per cent of the total number of adult females in a population that nest successfully, however, has not been heretofore satisfactorily determined for any population of wild geese. Only the number of pairs that *attempted* nesting has sometimes been known, but the *potential* number of breeding pairs available, based on complete knowledge of the age and sex structure of the population, has not been previously determined in conjunction with the subsequent reproductive performance of this entire component.

Recently, however, a breakthrough has been achieved in obtaining flyway-wide population data on the percentage of sexually mature females that nest successfully. These findings will be discussed in detail elsewhere (Hanson and Lumsden, *unpublished*), but because an attempt has been made here to provide a fairly complete discussion of the population dynamics of Canada geese (life-table aspects excepted), it was deemed desirable to round out these discussions by briefly summarizing a method of calculating the percentage of adult females in a population that raise one or more young.

The average brood size of the Canada geese (*interior*) of the Mississippi Valley Flyway in northern Ontario has been determined annually since 1959 by means of aerial photographs (Hanson and Lumsden, *unpublished*). We were concerned, however, by the fact that we did not know to what extent annual variations in the percentage of adult geese that successfully nested affected the over-all reproductive effort of the population. A record drop in the productivity of the Mississippi Valley Flyway population in 1963 stimulated a more careful consideration of our data; subsequently, it was realized that a few simple calculations could provide a more accurate assessment of over-all breeding success of the total number of females of breeding age in the population than that which is likely to be obtained from any other kind of field data. The following steps are involved (symbols in parentheses apply to the formula below):

1] The number of immatures in the flyway that left the breeding grounds is determined by multiplying the per cent of immatures in the fall or winter trap samples (Iw) by the total flyway population (N). If these data relate to the midwinter period, the calculated numbers of immatures must be corrected by adding the mortalities (Mh) incurred from hunting and the estimates of natural mortality (Mn).

2] The number of *productive* females (Fp) (or pairs) is determined by dividing the number of immatures calculated to have left the breeding grounds by the average brood size (B).

3] The total number of potential nesting females (Fn) in the flyway population is determined by multiplying the per cent of adult females (females 2 or more years of age) in fall or winter trap samples (Fw) by the flyway population. (For comparative purposes — and possibly a more accurate calculation of the number of breeding females — figures for the combined per cent of yearlings and adult females in the population and the total population figure for the winter previous to the nesting season could be used.)

4] The *per cent* of successful females (Fs) is then calculated by dividing the number of productive females by the total number of adult females calculated to be in the flyway. Hence:

$$\frac{\frac{(Iw \times N) + Mh + Mn}{B}}{(Fw \times N) + Mh + Mn} = \frac{Fp}{Ft} = Fs$$

The interrelationships between environmental forces and biological factors in determining the per cent of females that nest successfully and the size of the broods they raise can be fully understood only from studies of the entire reproductive cycle. Nevertheless, it is also important to carry out the above kind of assessment of the breeding season so that the depressive factors in a given year can be more effectively sorted out; once this has been done, the *whys* and *hows* of what has occurred can be more efficiently studied. The above calculations also provide the advantage of a flyway-wide evaluation of the breeding cycle without engaging in a time consuming and costly nesting study. Data obtained for 5 years for the Mississippi Valley Flyway population show that the variations in the per cent of females in the population that produce young from year to year are usually a much more important factor in determining total recruitment than are annual variations in the average size of broods. For example, in 1963, the per cent of adult females that produced young was one-third to one-fourth that of normal; yet average brood size that year was normal (Hanson and Lumsden, *unpublished*).

Regulatory Mechanisms

The ultimate objective in studying an animal population is to gain a greater refinement of understanding of how its numbers are adjusted in relation to its total ecology. In view of the numerous nesting and behavior studies that have been made of the large races of Canada geese, it would seem desirable to conclude this chapter on population dynamics with a brief discussion of the modes of population adjustment in Canada geese. In doing so, it is hard to escape making merely a modified restatement of the concepts advanced by Errington (1962) on the factors involved in the regulation of vertebrate populations.

The upper limits of populations of *maxima* and *moffitti* appear to be determined by the quality of the nesting habitat expressed in terms of the number of nest sites available in a suitable environmental complex. During my studies of goose breeding habitats in the muskeg of northern Ontario, my Indian guide, Joseph Chokomoolin, stated that "the geese may nest almost anywhere," but my observations revealed that nest sites fell into a number of clear-cut categories. The carrying capacity of an area is determined to a large degree by the number of attractive nesting niches available, but if these are already occupied by breeding pairs, then territorial conflict becomes the dominant factor in regulating the population through several or all of the following mechanisms: harassment of subdominant pairs by dominant pairs resulting in the prevention of normal courtship or breeding activities, eviction of subdominant pairs from the better nest sites, deferment of breeding by 2-year-old geese and some subdominant older pairs that might have otherwise nested, increased number of dropped eggs, increased rates of infertility, higher rates of embryonic mortality, and nesting in inferior habitat where greater exposure to the various agents of nest destruction results in higher nest losses.

The question arose as to whether an internal adjustment of productivity to population size (nesting pairs) was indicated by statistics for the Waubay flock. Banko (1960) has convincingly demonstrated this relationship for the trumpeter swan populations in Montana. The data for the Waubay flock, summarized in Table 27, indicate that giant Canada

27. Nesting success and average number of young produced per pair in relation to number of pairs present at the beginning of the nesting season, Waubay National Wildlife Refuge, 1945–62

Productivity factors	Number of pairs[1]							
	10–14 (3 years)		15–19 (7 years)		20–24 (5 years)		25–29 (2 years)	
	Average	Range	Average	Range	Average	Range	Average	Range
Nesting success	75.7	64–83	63.3	27–80	65.0	50–75	60.0	54–66
Young produced per pair[2]	4.1	3.1–4.8	2.7	0.6–4.1	3.3	1.9–4.5	2.9	2.7–3.1

1. Includes both successful and unsuccessful pairs. 2. In 1950 there were reported to be 68 pairs; of these only 33.8 per cent were successful and an average of only 1.6 goslings were produced per breeding pair.

goose populations, like trumpeter swan populations, behave according to Errington's "principle of inversity" – the inverse relation of the annual productivity of a population to its size. There is considerable variation in the Waubay data, but variation in environmental factors would be sufficient in some years to mask out internal adjustments of the population.

According to Gibb (1961), density-dependent fecundity has been indicated in birds only in the genus *Parus*. Studies of three European species have shown that clutch size is slightly smaller and fewer second nestings are attempted when populations are high. Gibb has attributed the inverse relationship between the size of spring populations and reproductive gains exhibited by gallinaceous birds to density-dependent mortality among the young. Obviously the most salient aspects of reproductive behavior of the Waubay flock (Table 27) fit neither category. Unless the age structure of the breeding adults changes, average clutch size tends to remain constant; also, there are no reports to indicate that gosling mortality is appreciably higher when broods are large than when they are small, although it is reasonable to suspect some difference may be eventually found. What the Waubay data do seemingly indicate is a type of population adjustment midway between the two listed by Gibb: a density-dependent productivity that is expressed by the degree of nesting and hatching success achieved (Table 27).

Except in atypical or peripheral areas of the nesting range — as at the Seney National Wildlife Refuge — it is unlikely that food is normally a limiting factor on the breeding grounds. On some wintering grounds, however, either or both the quantitative and qualitative aspects of the food resource may be important in limiting population size. The size of the available food supply was probably more significant before the advent of cultivated crops than now, as it is probable that in primitive times the giant Canadas that wintered on the north-central prairies were found in small, scattered concentrations, chiefly where spring-fed streams were present. Even under these circumstances, food crises were likely limited to periods of extreme cold and deep snows. For example, in 1963, most of the geese in the Rochester flock became dangerously low in weight following about 6 weeks of intense cold and deep snow, but few if any deaths could be attributed to the weather that year. At the Waubay National Wildlife Refuge, however, before the local breeding population established a migratory tradition, there were notable losses on two occasions due to blizzards: in 1942, 35 pinioned geese were lost; and from January 14–16, 1948, 26 free-flying geese died. After January 1955, the Waubay flock no longer attempted to winter in the area. Severe weather (winter storms, late spring cold, and summer droughts) may temporarily reduce a population from time to time, and cyclic changes in the climate may affect the habitat to the extent that a "new set of ground rules" in terms of an altered environment may prevail, but these factors do not regulate Canada goose populations in the connotation of the present discussion.

Another factor that *may* have a regulatory influence on the size of Canada goose populations is the effect of crowding on the wintering grounds. It is postulated that the visual stimulus of massed flocks and the accelerated social conflicts associated with crowding, especially in winters following breeding seasons in which a high per cent of the females were successful, have a depressive feedback on the endocrine system which, in turn, suppresses reproduction the following spring—particularly in respect to lowering the per cent of females that nest *successfully* (Hanson and Lumsden, *unpublished*).

This opinion is further based on the history of the Canada goose populations at the three main refuges in southern Illinois: Horseshoe Lake, Alexander County; Crab Orchard National Wildlife Refuge, Williamson County; and the Union County State Refuge. All are within 38 air miles from each other and the Horseshoe Lake and Union County refuges are only 15 air miles apart. The flock at the Horseshoe Lake refuge originated from the geese that wintered on adjacent sectors of the Mississippi River; the Crab Orchard and Union County refuge flocks were, in turn, from birds en route to Horseshoe Lake, but at both of these refuges the pioneering flocks built up independently and have since adhered to their own traditional wintering area. They have also tended to stabilize independently despite their sharing a common breeding ground which all available evidence indicates could support a great many more geese than it does now. It would appear that, contrary to what is the seemingly underpopulated status of the refuges in winter from the standpoint of unused space and food available, the numbers of geese at these refuges may be oscillating around a saturation level. Hunting in recent years is not believed to be a primary factor as a quota system has held kills to levels which should have permitted these populations to have increased annually.

Parasites are seldom a cause of mortality in adult Canada geese unless they are simultaneously under stress from other factors; they are, however, an occasional cause of mortality in goslings. An apparent relationship between a winter diet low in protein and parasite infections has been described (Hanson and Gilford, 1961). Nevertheless, the possibility of a reverse situation should not be overlooked; heavily infected individuals may be debilitated and thereby placed at a competitive disadvantage.

Research of recent years has increasingly revealed the degree to which populations of many animal species are self-regulatory. The present state of knowledge on Canada geese supports this concept; final limitation of numbers in this species appears to reside in internal controls.

NOTE TO CHAPTER 13

1 However, a study currently being carried out on the productivity and population dynamics of *interior* indicates annual alternations of high and low productivity of similar magnitude (Hanson and Lumsden, *unpublished*).

Management 14

COMPETENT management of wildlife populations requires an understanding of ecological relationships that are sometimes of a fairly subtle nature. Fortunately, much of the ecology of Canada geese is uncomplicated; their basic requisites are known and readily met. What is of chief concern here are some of the distinctive attributes of *maxima* populations which must be taken into consideration in formulating management plans for this race.

Taxonomy

It is of paramount importance in carrying out management programs and banding operations that stocks of *maxima* be distinguished from other races, particularly from *moffitti, interior,* and an undescribed race which is smaller than *maxima* but similarly colored (*see* note 1, "Discussion"). The detailed and comparative descriptions of *maxima* given earlier should now make this task relatively easy. Clearly, banding operations on a refuge where appreciable numbers of *maxima* winter together with one of the other large races will have only limited value unless the racial identity of each goose banded is determined.

There have been various attempts to establish breeding populations of Canada geese on national refuges that lie in the former range of the giant Canada goose. In some instances, stocks of *maxima* were used; in other attempts, stocks of *moffitti* and *interior* were released. In 1962, Harvey K. Nelson and I reviewed these projects in the field, and we found that only stocks of *maxima* reproduced and became permanently established (Nelson, 1963). Now that adequate seed stocks of *maxima* are known to exist, it is strongly recommended that all current breeding stocks on refuges within the original range of the race be carefully screened and that flocks not meeting the subspecific standards of *maxima* be eliminated and replaced by the latter. It is perhaps redundant to reiterate the axiom that the plants and animals of a region are the product of long evolutionary development; transplanted races which evolved under different environmental conditions cannot be expected to be as successful in a new area as the original stock which was adapted to it.

Behavior

The behavior traits of the giant Canada goose are both its strength and its weakness as a game bird. Partly because of its placid disposition, it readily accepts the proximity and protection of man. Perhaps its acceptance of man's activities also relates, paradoxically, to its having a greater degree of intelligence than the other races; its large head size — hence, brain size — (Fig. 18) makes this assumption tenable (*see* Rensch, 1956, 1959).

The rapidity with which a flock of giant Canada geese will become semidomestic has certain advantages in the management of locally breeding flocks. A flock that has been fed most of the year at an established site in a refuge area may scatter widely over the surrounding countryside to nest but will later return to it with their young. This behavior trait has been shown by flocks at the Delta Waterfowl Research Station, the East Meadow Ranch, Oak Point, and at the Alf Hole refuge, Rennie, all in Manitoba; the Round Lake Waterfowl Station, Minnesota; and at the Des Plaines Wildlife Area, Illinois. The first three flocks are free-flying and migratory; hence, this behavior is not a result of domestication, but it may be exploited to protect local breeding stock from hunting.

The giant Canada geese tend to fly lower than the other races, particularly in making local flights. This trait was repeatedly noted by early observers of the race in North Dakota. It was, therefore, of interest to learn that the 19-pound goose killed at Horseshoe Lake in 1941, was only 5 or 6 feet off the ground when shot (William Brown, *personal communication*). Commenting on the 21-pound giant Canada that he shot on the Missouri River near St. Joseph, Missouri, Robert A. Brown (*letter*, July 8, 1963) wrote: "This particular goose came in by itself from up river. It came in quite low over the water, wheeled over our decoys, and started to settle in the water." Low flying would, of course, increase the vulnerability of a flock of *maxima* to hunting, a factor which should be considered in setting local regulations

Habitat and Population Management

In the north-central states, management of giant Canada geese falls into two categories: (1) the management of migrant populations on wintering refuges, and (2) management of refuge lands so as to achieve the greatest productivity of breeding stocks. Each type of population and refuge has its own inherent problems although some refuges may serve both migrant and resident populations. Refuge planning in either case must ultimately take into consideration the factors limiting these populations. In the brief introduction to this chaper, I stressed the *seeming* simplicity of the ecological relationships involved in Canada goose management; in reality, of course, biological relationships, whether internal or external to an animal, are never simple. Overriding the apparent ease of managing the land for Canada geese is the complexity of the responses of populations to their total environment and, perhaps most importantly, to their own numbers.

Perhaps the principal lesson we have learned from experience with refuge management in southern Illinois is that, if we are to continue to increase total flyway numbers, we must enlarge — perhaps more importantly create — new wintering areas and thereby build up and add new, independent flocks to the flyway population. Transferring this experience to wintering populations of *maxima*, the desirability of creating new wintering refuges in Minnesota and Wisconsin is apparent if we are to expand the populations breeding in Manitoba. However, as noted earlier, the wintering flock in Rock County has been static for many years. This would appear to be due to the marginal nature of their breeding grounds in eastern Manitoba which may preclude further increases, rather than to losses from hunting which, in southeastern Wisconsin, are known to be negligible.

Provisions for managing wintering areas for migrant populations of *maxima* from Canada differ in some respects from the problems met in managing wintering flocks of other races in more southerly areas. The availability of open water — even if limited to a spring-fed creek — may predetermine the locations at which refuges may be established. Large lakes for roosting are not a requisite as these geese are amenable to roosting on the ground. However, clipping the vegetation of such areas or having them grazed by cattle will make them more acceptable for feeding as well as for roosting. The flock of giant Canada geese that winters in Rock County, Wisconsin, roosts on the closely grazed bottomlands of Turtle Creek (Fig. 44).

I consider it to be extremely unlikely that the amount of available food in midwinter influences the productivity of goose flocks the following spring. A discussion of this viewpoint and suggestions for improving the quantity and quality of foods on refuges for wintering Canada geese have been presented elsewhere (Hanson, 1962).

The over-all size of populations that breed in the temperate zone (*maxima* and *moffitti*), where desert, mountains, and agricultural lands occupy so large a part of their general range, is restricted by the relative scarcity of nesting habitat and by competition for nest sites. With these populations there is the opportunity to increase local flocks through management practices, but, obviously, there are upper thresholds beyond which manipulations of the breeding habitat may be unable to bring about higher reproductive success. If inter-pair strife takes place during the early part of the breeding season to the extent that productivity is reduced below that normally expected, it can be assumed that the habitat *as it exists* has already reached its saturation point or is overcrowded. Manipulating the environment to provide more potential nesting territories and nest sites would then be in order. In evaluating the reproductive performance of a population of Canada geese, however, it would be well to bear in mind Kalmbach's (1939:596–97) lucid assessment of nesting success in waterfowl populations:

> It avails little to lament or endeavor to improve an apparently low degree of nesting efficiency if that degree is the normal for the species living under reasonably satisfactory conditions. We are not meeting our responsibilities on the other hand, if we are willing to rate as good enough a standard of egg hatch that field studies indicate is below normal and below which might be attained if management practice were devised to circumvent nesting hazards. Obviously it behooves us to learn what is the normal and, later, how far that degree of success may be improved by management.

An intensive restoration program must necessarily be based on federal, state, and provincial refuges which can provide the initial elements of protection for local breeding populations. The total acreage of present refuges may seem impressive but, relative to the problem of wide-scale restoration, it is limited and must, therefore, be managed with maximum efficiency. I was impressed with the effort made toward achieving this goal at the Seney National Wildlife Refuge in the Upper Peninsula of Michigan. Although a vast system of dikes created many pools and numerous natural islands on this refuge, the population has now leveled off at what would seem to be far below the potential nesting capacity of the area. Many pools and islands are not used by breeding pairs. The reason may be, in part, adherence to traditional nest sites by the adults, and to birth sites by the nonbreeding adults, but it may also be related to more complex ecological factors. There is a marked dearth of open, level feeding areas for the young broods at this refuge. Grazing areas are available only as scattered patches on the dikes where the families are exposed to predation. Perhaps a principal management objective on this refuge should now be to bring about a greater dispersal of nesting pairs coupled with the development of extensive and accessible grazing areas, preferably of bluegrass. It might be advisable to drain some of the pools and establish pastures in them or to develop more grassland areas adjacent to the pools wherever possible. At the present time, personnel on federal and state refuges are following the lead of George C. Arthur of the Illinois Conservation Department and are giving greater recognition to the need for a better balance between browse crops, such as native grasses and clovers, and the cereal grains on the larger managed units which are used both for wintering geese and as stopover points for migrating geese.

As stated earlier under "nest sites," muskrat houses have always been a key component of the nesting habitat of the giant Canada goose. In lieu of these, over-water nesting platforms can be substituted to insure safe nesting or a better dispersement of nests. Such structures have been well accepted by the flocks at the Agassiz and Waubay National Wildlife refuges (Fig. 70). At the East Meadow Ranch on Lake Manitoba, nesting success was poor until wood platforms were supplied for nest sites (Fig. 71). To encourage the use of these platforms at the East Meadow Ranch the first year they were set out, hay was stacked on the ice around their bases in early spring and also placed on the platform for nest material. This procedure permitted the mated pairs — which arrive before the breakup of the ice — to walk up onto these platforms when taking possession of them, thereby avoiding their initial reluctance to fly to them.

In 1944, a sportsmen's group in the Okanagan Valley of Washington capitalized on the tendency of Canada geese (*moffitti*) to nest in trees on the old nests of herons and hawks by placing basket-like platforms in trees. The acceptance of these platforms for nesting subsequently led to the trial and use of galvanized washtubs as nesting platforms (Yocum, 1952 and 1956). At the Trimble Wildlife Area, Missouri, washtubs placed in trees are now being routinely used to provide safe nesting sites for a population of giant Canada geese (Fig. 72).

MANAGEMENT POTENTIAL

The ready acceptance of nesting giant Canada geese to joint occupation of the land with humans and their tolerance of human activity are indicative of the great potential that exists for increasing the over-all population of this race. An impressive example of establishing a population of this race in an area of dense human population has been described by Karl Bednarik (*letter*, November 22, 1963):

> We are realizing our greatest gosling production on Lake St. Mary's. This artificial lake was created in the years 1837–1845 by damming the headwaters of the Wabash and St. Mary's rivers resulting in flooding the dividing area in southwestern Ohio between the rivers. The lake lies in a region with a resident human population level of 237 per square mile. The ten mile long lake has some 65 boat landings (bait stores, small boat harbors) around its periphery. Over 3,000 boat licenses for row boats, sail boats, and inboard motors are sold annually on this lake. The lake is heavily used by fishermen. In a recent 5-year period, anglers from all of Ohio's 88 counties, from 28 other states, and from two foreign countries were found angling on this lake. The shore line has been well developed with cottages.
>
> Our geese nest on the islands, along the shore-line, near or beside summer cottages, just about any place that you can imagine. The human residents around the lake take great pride in having free-flying geese nesting on their property. They 'turn-in' any individual molesting nesting geese to the local game warden.

ESTABLISHMENT OF BREEDING FLOCKS

Perhaps the principal question that will arise in an extensive restoration problem is — How can nesting populations of giant Canada geese be established on a new area? Fortunately, the experiences of H. Albert Hochbaum of the Delta Waterfowl Research Station have provided a workable technique. With the cooperation of the Manitoba Department of Mines and Natural Resources, he was instrumental in the successful re-establishment of the giant Canada on the Delta marshes and the marshes of the East Meadow Ranch. At both areas, the transplant stocks of immatures were held in captivity until they were 2 years old. By that time, these geese had accepted the local area as their traditional home breeding range, an adoption which was doubtless reinforced by the fact that their sexual maturation occurred on the intended breeding grounds.

The re-establishment of *maxima* in presently unused areas of its former range offers conservation-minded groups unusually suitable projects; however, the success of such projects would necessarily depend on the full support of the local community. It would, for example, scarcely be the "inalienable right" of local hunters to harass a goose population that would be present in their area only by virtue of the dedicated efforts of a comparatively few individuals. Once a population of giant Canada geese is restored to an area it will afford some hunting opportunities, but it should first be encouraged for its aesthetic values and as a signal wildlife remnant of earlier times.

Many communities in Minnesota are currently starting flocks of giant Canadas on lakes lying within their towns or adjacent to them. Some of these lakes provide the advantage (as in the case of Silver Lake at Rochester) of remaining partly open in winter because their waters are used for cooling purposes by the town power plant. Those personnel considering a restoration program should keep in mind the following specific points as to the behavior traits of *maxima* populations and some of the problems and costs entailed in supporting a community refuge and a breeding flock.

1] Under climatic conditions no more severe than those prevailing in winter in southern Minnesota, free-flying birds will not migrate if open water and food are locally available. These factors may necessitate a feeding program to protect them from local hunting.
2] If the flock migrates, it will reassemble in the spring on the home area on which the original stock was released and fed. Soon afterwards, the mated pairs will disperse to the surrounding countryside to nest, but after the young have hatched the family units will again attempt to return to the home area. As the entire family will be flightless at this time, there may be considerable mortality, particularly if there is mesh wire fencing between the nest site and the headquarters which will prevent the overland return journey.
3] Protection from local hunting will be a primary problem. If a town lake is used principally for roosting, it will be necessary to establish a closed peripheral buffer area and a closed feeding area. A 2-square-mile area of farmland has proved to be remarkably effective in protecting the Rock County flock which usually numbers between 2,000 and 5,000 geese.
4] Considerable costs may be involved in the initial stages of the program. In addition, a winter feeding program and the leasing of land for refuge purposes will entail reoccurring costs.
5] State and federal permits will be required.

INVENTORIES AND HUNTING REGULATIONS

Reliable statistics on the populations and kills of the giant Canada goose should be obtained annually; however, these data would be difficult to collect over the widespread range of *maxima*, especially in areas where populations of *maxima* become intermixed with *interior* and *moffitti*. To date, reasonably adequate flyway data of these kinds are available and are continuing to be collected only for the race *interior* (Hanson and Smith, 1950; Hanson and Currie, 1957).

Adjustment of the opening and closing dates of the hunting season is one of the principal means of regulating the size and selectivity of waterfowl kills. It should, therefore, again be pointed out that because the fall migration movements of the giant Canadas are fully a month behind the migration of the race *interior*, late hunting seasons will tend to jeopardize stocks of *maxima* more than will early seasons. The giant Canada geese in southern Manitoba do not begin to leave until the last third of October, and in some years large numbers remain in the province well into late November, by which time the concentrations of *interior* in Wisconsin and Illinois are nearing peak numbers (*see* Hanson and Smith, 1950:111).

Discussion 15

THE UNDERLYING thesis of this book is that the present-day stocks of large Canada geese that now nest in the midsector of the continent east of the Rocky Mountains constitute a single recognizable race and are in no way distinguishable from the stocks that bred in the region at the time of early settlement. It may be difficult for a few readers to accept this premise in view of the seemingly appreciable size differential between some known populations, but the total evidence available supports this concept.[1]

In part, much of the fame of this race has accrued from the weights recorded for a comparatively few very large males. Some of the factors that explain the occurrence of these excessively large individuals have been discussed in Chapter 2, "Physical Characteristics," but the principal environmental influence determining the average size of all subspecific populations of Canada geese is climate (*see* footnote 2, "Breeding Range"). The original range of *maxima* extended from Reelfoot Lake, Tennessee, to the region of Edmonton, Alberta, a NW–SE distance of 1,500 miles, and in terms of latitude, 1,090 miles. Although this race is primarily adapted to the prairie biome, no other subspecies of wild goose in North America nests over so large an area or one which includes such a diversity of habitats and climates. Climatic data for some stations near known nesting localities are given in Table 28. Closely associated with the length of the growing season for plants is the length of time that water areas remain open, especially the period after the goslings have hatched (particularly relevant for races in Arctic areas). Directly and indirectly these factors effectively limit the period of growth for the young and the time available for the wing molt of the breeding adults following nesting. Because climatic factors on the breeding range

28. Average July temperatures, average inclusive dates of frost-free period, and number of frost-free days at selected localities in the breeding range of the giant Canada goose

Locality	Average July temperature	Average inclusive dates of frost-free period	Number of frost-free days
Calgary, Alberta	61.9	May 28–Sept. 15	111
Saskatoon, Saskatchewan	65.7	May 30–Sept. 12	106
The Pas, Manitoba	64.5	May 30–Sept. 21	115
Portage la Prairie, Manitoba	69.2	May 22–Sept. 25	127
Malta, Montana	70.4	May 14–Sept. 24	133
Devils Lake, North Dakota	68.0	May 15–Sept. 23	131
Thief River Falls, Minnesota	68.0	May 25–Sept. 17	115
Fairmont, Minnesota	73.3	May 5–Oct. 5	153
Racine, Wisconsin	72.0	Apr. 24–Oct. 21	180
Denver, Colorado	72.5	Apr. 26–Oct. 14	171
Kankakee, Illinois	75.4	Apr. 30–Oct. 14	167
Cairo, Illinois	79.7	Mar. 29–Nov. 1	217

and physical size have shown good *intersubspecific* correlation, it is reasonable to conclude that climatic factors are also operative *intra-subspecifically* as evolutionary forces and that, in a race having as great a range latitudinally as *maxima,* a cline in size can be expected.

The morphometric data presented in this paper have dealt with what at first might seem to be two distinct subpopulations of *maxima* of different size: the migrant Canadian stocks wintering in southern Minnesota, and the captive and wild flocks in North Dakota, Minnesota, and Illinois, which are nearly all derived from old decoy flocks. These data give rise to three basic questions: (1) Are present-day stocks of *maxima* from the northern plains states similar in size to wild stocks formerly in that region? (2) Are the geese nesting on the northern fringes of the range in Canada smaller than the stocks from southern Manitoba? (3) Were the geese that formerly nested at the extreme southern part of the range even larger than any of our existing stocks? I believe that the answer to the first two questions is "yes" and, on the basis of the osteological record, it is "no" to the third.

Skins of Canada geese collected prior to 1900 in the north-central states during the breeding season and all early records of weights and wing spans do not differ significantly from those available for present-day refuge and captive stocks. In fact, there is reason to believe that some private game breeders have selected for large size in maintaining their breeding stock. Bob Elgas, a noted game bird breeder in Big Timber, Montana, has concurred with this view (*personal communication*).

Two fossil bones of *Branta canadensis* from Minnesota indicate the minimum age of this race (I believe it to be one of the oldest races, predating the Pleistocene period) and provide some insight into its distribution and size in Pleistocene times. I am quoting Wetmore's (1958:6–7) discussion of one of these bones, a portion of an ulna, in its entirety because of the important support it gives to the basic thesis of this book.

> The Canada Goose, *Branta canadensis* (Linnaeus), distributed across the continent in the present day, seems to have had equally wide range for a long period of time, since its bones have been reported from Pleistocene deposits in Oregon, California (including offshore Santa Rosa Island), and Florida, and from beds of supposed Pleistocene age in Nevada. As an additional, interior, link between the western and the southeastern localities it is of interest to report the occurrence of the species at St. Paul, Minn. The record is based on the distal end of a right ulna, sent to me for identification by Scott K. Wright of that city. Mr. Wright reports that the bone was found at the bottom of a large trench dug by the City Water Department in an ancient peat bog. Bones of the Pleistocene *Bison occidentalis* came from the same trench, though it is noted that there were also other remains identified as the modern *Bison bison*. While the goose bone was not encountered in place, having been found, as stated, in the bottom of the trench, Mr. Wright believes it to be of Pleistocene age. This conclusion is substantiated by the condition of the bone, which has lost all free animal matter, in addition to having the dark brown discoloration usual to specimens found in peat deposits. The occurrence of *Bison bison* should not preclude a late Pleistocene age so that it appears proper to record the occurrence as Pleistocene.

A fossil femur, which was found under circumstances similar to those related above and which was similar in color and general appearance to the ulna fragment described by Wetmore, was made available to me for study by Dwain W. Warner of the Minnesota Natural History Museum. I assume that both specimens came from the same skeleton. The length of this femur is identical to the largest size present-day examples of *maxima* and exceeds the upper range in size of *interior* (Table 29).

Early in this study I anticipated that bones of giant Canada geese could be found in osteological collections from the kitchen middens of prehistoric Indians. Subsequently, I was gratified when Harry G. Lumsden called my attention to the recovery of a humerus of a Canada goose from a 1,300-year-old Indian site near Port Maitland, Ontario (cited previously under the range of *maxima* in Ontario). The length of this humerus, now in the Royal Ontario Museum, identifies it as being from a giant Canada goose (Table 29).

When inspecting the collection of bird bones from Indian sites at the Illinois State Museum with Dr. Paul W. Parmalee, I noted an incomplete humerus of a Canada goose in a collection made in 1962 from the 2,000-year-old Snyder site in Calhoun County, Illinois. It was of such great size that it immediately attracted my attention. Comparative studies showed that it dwarfed humeri from *interior*, as well as those available at the time from several populations of *maxima*. Rather it proved to be similar in size to the humerus of a whistling swan and quite closely resembled its qualitative aspects. Therefore, I assumed that if this large humerus were truly from a giant Canada goose, it would likely be equaled in size only by a humerus from a 16 to 18-pound bird. Fortunately, near the close of this study, I was able to almost match this large humerus from Illinois with a humerus dissected from one of the three 12-pound specimens of giant Canada geese collected at Denver, Colorado (Fig. 73).

The Canada geese from the northeastern portion of the range of *maxima* in Manitoba are said to be smaller than the southern Manitoba stocks (H. Albert Hochbaum, *personal communication*). This reported relationship is in agreement with the N–S clinal variation in size found in Canada geese and is also suggested by the marked variation in size found each winter in the wintering flock at Rochester, Minnesota, which is derived from Manitoba.

The continuity of expression from flock to flock, both in the wild and in captivity, of distinctive characters — white head spots, prominent neck rings, types of cheek patches, spatula-shaped bills, unique scaling of the tarsi, and long neck — that typify the race is impressive evidence of common genetic background. Should it seem desirable to further subdivide this subspecies, both for better understanding of the biology of the species and for practical advantages in management, where should the range be divided? Surely not along the invisible international boundary between the United States and Canada which Delacour inadvertently implied by extending the range of *moffitti* to southwestern Manitoba while including all of North Dakota in the former range of *maxima*. It is an inescapable fact that the prairies of the north-central states and the prairie provinces of Canada form an immense continuum of habitat without important ecological or physical barriers which so conspicuously separate the breeding ranges of nearly all of the other races of Canada geese.

The various populations of *maxima* differ considerably in coloration as well as in size. Captive flocks in Illinois, at Round Lake, Minnesota,

29. Lengths in millimeters of some long bones of fossil and recent specimens of the genus *Branta*

Bone and species or subspecies	Locality
HUMERUS	
B. c. maxima	Calhoun Co., Illinois
B. c. maxima	Port Maitland, Ontario
B. c. maxima	Denver, Colorado
B. c. maxima	Round Lake, Minnesota
B. c. maxima	Rochester, Minnesota
B. c. interior	Horseshoe Lake, Illinois
FEMUR	
B. c. maxima	St. Paul, Minnesota
B. c. maxima	Denver, Colorado
B. c. maxima	Round Lake, Minnesota
B. c. interior	Horseshoe Lake, Illinois
TIBIOTARSUS	
B. dickeyi	McKittrick, California
B. c. maxima	Denver, Colorado
B. c. maxima	Round Lake, Minnesota
B. c. interior	Horseshoe Lake, Illinois

and the wild population at Denver, Colorado, are relatively dark and richly colored compared with wild stocks in Manitoba which are more grayish and less brownish in color (*see* Chapter 2, "Physical Characteristics"). The geographical origin of most captive stocks, many of which are descendent from old decoy flocks that were formerly widely sold and shipped about, is not precisely known; consequently, it is not possible to assemble a series of specimens to document ideally a clinal east to west — and possibly also a south to north — variation in coloration, but existing stocks do appear to indicate a cline that conforms with Gloger's rule. Although the dark-colored stocks on the whole tend to be the largest, some of the largest geese that have been shot have been light-colored birds (Figs. 7 and 8).

The present findings on the variation between stocks of *maxima* are altogether in agreement with findings for other species of birds. It is of common occurrence for a race to exhibit combinations of characteristics of adjacent races (E. R. Blake and A. L. Rand, *personal communications*) — *moffitti* and *interior*, in the case of *maxima*. And, as Mayr (1951:94) has pointed out, "much geographical variation, particularly on continents, is clinal. These character clines are correlated with gradients in the selective factors of the environment." What Mayr (1963:

Geological or historical age of specimen	Age and sex class of specimens	Number	Average length	Range	Authority
2000 B.P.	?	1	(see Fig. 80)	—	This paper
1300 B.P.	?	1	201	—	This paper
1963	Adult male	3	202.7 ± 3.3	198–209	This paper
1961–63	Adult male	1	190	—	This paper
1962	Adult male	1	193	—	This paper
1950–53	Adult male	25	181 ± 1.3	166–192	Hanson and Fisher, *unpublished*
Early(?) post glacial	?	1	92.0	—	This paper
1963	Adult male	3	91.3 ± 0.7	90–92	This paper
1962	Adult male	2	91.0 ± 1.0	90–92	This paper
1950–53	Adult male	25	84.2 ± 0.6	76.9–91.9	Hanson and Fisher, *unpublished*
"McKittrick Pleistocene"	?	1	177.0	—	Miller (1924)
1963	Adult male	3	159.0 ± 0.6	158–160	This paper
1962	Adult male	2	163.0 ± 3.0	160–166	This paper
1950–53	Adult male	25	158 ± 1.1	145–170	Hanson and Fisher, *unpublished*

247–48) has further emphasized to be true for specific variation has been demonstrated here to hold true for subspecific variation: "The more closely a species is studied, the more likely it is that some evidence will be found for ecological polymorphism or gradual ecological variation. This variation is still largely ignored by ecologists, most of whom discuss the ecological requirements of a species in a strictly typological manner."

The clinal nature of the variation in size and coloration in *maxima* could doubtless be more fully documented *if* there were adequate early collections from over all of the original range and *if* existing captive stocks had not been widely transplanted and their blood lines intermixed. I have given considerable stress to the ecologically related attributes of *maxima* not only for their own interest but for their supportive taxonomic values. In doing so, I have brought this study full circle and, I trust, successfully exemplified Mayr's (1950:123) opinion that "the study of the role of ecological factors has become one of the most important objectives of taxonomy."

Presuming that the question of the present-day existence of the race *maxima* has been satisfactorily answered, the corollary question, How large is the present-day population? remains to be answered. In Tables 30 and 31, I have summarized the numerical status of the race. The data

30. Approximate numbers of giant Canada geese in Canada and United States as of 1962–1963, exclusive of stocks held under United States Bureau of Sport Fisheries and Wildlife permits

Province or state	Area	Number
Alberta	South of the Red Deer River	15–20,000[1]
	Elsewhere	3,000
Saskatchewan	U. S. Border to S. Sask. River	2,667
	S. Sask. River to N. Sask. River	1,000
	E. Sask. (Quill L., Kinistina, Tisdale, Yorkton areas)	1,667
Manitoba	Range in Province	15,000
Ontario	St. Lawrence Seaway Commission Park	300
	Niska Waterfowl Research Station	32
Montana	Hi-Line Unit	3,089
	East Slope Unit	795
	Helena Unit	624
Colorado	Denver Area	1,000
	Fort Collins Area	400
North Dakota	Lower Souris N. W. R.	5
	Upper Souris N. W. R.	160
	Lostwood N. W. R.	30
	Sullys Hill N. W. R.	22
	Snake Creek N. W. R.	68
South Dakota	Sand Lake N. W. R.	270
	Waubay N. W. R.	380
	Lacreek N. W. R.	125
Nebraska	Crescent Lake N. W. R.	200
Minnesota	Agassiz N. W. R.	550
	Tamarac N. W. R.	69
	Rice Lake N. W. R.	42
	Lac Qui Parle State Refuge	170
	Thief Lake State Refuge	200
	Carlos Avery State Refuge	150
Missouri	Trimble Wildlife Area	320
	Shell-Osage Wildlife Management Area	80
Wisconsin	Crex Meadows Wildlife Area	164
	Bay Beach Wildlife Area	100
	Barkhausen Game Sanctuary	200
Michigan	Seney N. W. R.	1,100
	Shiawassee N. W. R.	400
Illinois	Des Plaines State Game Farm	385
Indiana	Jasper-Pulaski State Game Preserve	50
	Tri-county State Fish and Game Preserve	22
	Lake Sullivan	50
Ohio	Lake St. Mary	1,100
	Mosquito Creek	600
	Kildeer Plains	500
Total number (approximate)		54,586

1. 17,500 used in arriving at total.

31. Numbers of Canada geese of all subspecies[1] held by game breeders under United States Bureau of Sport Fisheries and Wildlife permits as of January 31, 1963[2]

State	Permittees	Number of geese held 0–10	11–20	21–50	51–100	100+	Total Held
Ohio	85	70	11	3		1	761
Indiana	117	91	16	7	3		1,192
Illinois	291	187	52	42	9	1	3,667
Michigan	79	55	13	10	1		790
Wisconsin	170	126	20	16	7	1	1,909
Minnesota	145	98	24	15	7	1	1,642
Iowa	155	104	23	21	5	2	2,026
Missouri	65	37	13	12	3		932
North Dakota	22	14	4	2	2		315
South Dakota	35	20	8	5	2		488
Nebraska	78	53	17	7		1	859
Total	1,242	855	201	140	39	7	14,581

1. No distinction is made for subspecies. It is known some of the permittees have limited numbers of the smaller races (*minima, hutchinsii, parvipes,* etc.); many, however, have extremely large birds (*maxima*). 2. Data courtesy of Harvey K. Nelson and Division of Game Management.

in parentheses are partially arbitrary estimates (*examples:* "elsewhere in Alberta," and "elsewhere in Manitoba") or are derived figures based on the number of breeding pairs censused in an area (Saskatchewan). In this latter instance, I have assumed that the number of breeding pairs estimated to be in an area constitutes roughly 30 per cent of the endemic population.

To sum up, the data at hand in 1963 pointed to a population of about 55,000 wild giant Canada geese on government refuges and privately owned lands in Canada and the United States. In addition to this wild population, it can be reasonably assumed that of the 14,581 Canada geese of various races reported to be held by private individuals in the north-central states under federal permits, *at least* half of these are giant Canada geese. Manifestly, insofar as its survival is concerned, the race is in a secure position. This conclusion stimulates a third question, How well will the opportunity to expand and manage these populations be exploited? The outstanding programs that have been underway on the various state, provincial, and federal refuges for a number of years and the privately initiated programs in many communities provide assurance that the future of the giant Canada is indeed bright.

The giant Canada goose, newly emerged from obscurity, is again a part of our living heritage. To the casual reader it may appear that our present knowledge of this bird is already formidable; the student of avian biology, however, will realize that it is only a temporary platform of departure for more refined and perceptive studies.

NOTE TO CHAPTER 15

1 My studies of skins of *Branta canadensis* in various museums in the spring of 1963 indicated the existence of an undescribed race which was slightly smaller than *interior* but which in coloration and white markings superficially resembled *maxima*. Photographs which I have seen of geese shot on the Platte River in Nebraska (also three specimens in the Denver Museum of Natural History) and at Grande Prairie, Alberta, suggest that this race is the predominant migrant population across the prairies. A few individuals disperse in winter as far east as the Mississippi River. One example, an immature, was trapped at Horseshoe Lake in the fall of 1958. Because I was puzzled by its identity, I photographed it for later reference (Fig. 74).

A study with Harold H. Burgess in December, 1964, of 274 heads from Canada geese that had been killed by hunters near the Squaw Creek National Wildlife Refuge, Missouri, revealed that six (possibly seven) races of Canada geese migrate through or winter on this refuge. Classification of these heads as to subspecies indicated that the undescribed race mentioned above comprised at least 60 per cent of the kill and that *maxima* ranked second in the hunters' bag.

Basing judgment on the size and coloration of this new race and known clinal variations of the species, it was anticipated that it would be found to breed in northwestern Manitoba and northern Saskatchewan, a region which a few scanty reports (Buchanan, 1920; Randall, 1962; Alex Dzubin, *letter*, June 30, 1964) had indicated held a scattered population of Canada geese. However, I failed to realize this expectation when Eugene F. Bossenmaier, Harry G. Lumsden, and I collected four breeding adults in July, 1964, on Fidler Lake at the east end of the Churchill River basin in northern Manitoba. It is significant that the population of Canada geese in this area can be associated with a distinct type of habitat and is isolated from the population of *interior* nesting in the coastal muskeg west of Hudson Bay. Subsequent comparative studies of the four specimens collected revealed that they were similar to *interior* in size (and weight) and coloration but their bills were more pointed and slender and of a lower profile. One of these adults had been banded at Squaw Creek refuge. Field investigations to delineate the ranges of these new races, which will be formally described in a report under preparation, are continuing.

LITERATURE CITED

Agersborg, C. S. 1885. The birds of southeastern Dakota. Auk 2(3):276–289.

Aldrich, John W. 1946. Speciation in the white-cheeked geese. Wilson Bull. 58(2):94–103.

American Ornithologists' Union. 1957. Check-list of North American birds, 5th Ed. Publ. by the Union. 691 p.

Anderson, Rudolph Martin. 1907. The birds of Iowa. Proc. Davenport Acad. of Sci. 11:125–417.

Anonymous. 1874. The largest goose on record. The American Sportsman 4(9) (new series #35):139.

Audubon, Maria R. 1897. Audubon and his journals. The Missouri River journals. 1960. Dover Publications, Inc., N.Y. Vol. I:452–532, Vol. II: 3–195.

Austin, Oliver Luther, Jr. 1932. The birds of Newfoundland, Labrador. Memoirs of the Nuttall Ornithological Club. No. VII, Nuttall Ornithological Club, Cambridge, Mass. 229 p. Map at back.

Balham, Ronald W. 1954. The behavior of the Canada goose (*Branta canadensis*) in Manitoba. Ph.D. Thesis. Univ. of Missouri. 229 p.

Banko, Winston E. 1960. The trumpeter swan. Its history, habits and population in the United States. U.S. Dept. of Interior. North Amer. Fauna 63:1–214.

Barrows, Walter Bradford. 1912. Michigan bird life. Spec. Bull. Dept. Zool. & Physiol., Mich. Agric. College. 822 p.

Beebe, C. William, and L. S. Crandall. 1914. Specialization of tail down in certain ducks. Zoologica. 1(13):248–252.

Benedict, Francis G., and Robert C. Lee. 1937. Lipogenesis in the animal body, with special reference to the physiology of the goose. Carnegie Institution of Wash. Publ. No. 489. 232 p.

Bent, A. C. 1902. Nesting habits of the anatidae in North Dakota. Auk 19(2): 165–174.

Bent, Arthur Cleveland. 1925. Life histories of North American wild fowl. U.S. Nat. Mus. Bull. No. 130. 316 p.

Bird, Ralph D. 1961. Ecology of the aspen parkland of western Canada in relation to land use. Canada Dept. of Agric. Research Branch Publ. 1066. 155 p.

Boldt, Wilbur. 1961. North Dakota's vanished wildlife. North Dakota Outdoors [July]. Dakota territory centennial issue 10.

Boyd, Hugh. 1953. On encounters between wild white-fronted geese in winter flocks. Behavior 5(2):85–109.

Brakhage, George. 1962. Canada goose nesting and management studies on selected areas in Missouri. Federal Aid in Wildlife Restoration Act. Project No. 13–R–16. 34 p.

Braun, Lucy. 1950. The deciduous forests of eastern North America. Blakiston, Phila. 596 p.

Bruner, Lawrence. 1896. Some notes on Nebraska birds. (Reprint from the report of the Nebraska State Horticultural Society for the year 1896.) Lincoln. 178 p.

Bruner, Lawrence, Robert H. Wolcott, and Myron H. Swenk. 1904. A preliminary review of the birds of Nebraska with synopses. Kloppand Bartlett Co., Omaha, Nebraska.

Buchanan, Angus. 1920. Wildlife in Canada. John Murray, London. 264 p.

Burgess, G. C. 1957. Occurrence of *Leucocytozoon simondi* M. and L. in wild waterfowl in Saskatchewan and Manitoba. Jour. Wildl. Mgt. 21(1): 99–100.

Butler, Amos W. 1897. The birds of Indiana. p. 516–1187.

Canada Dept. of Mines and Resources. 1945b. Geological map of the Dominion of Canada. Map 820A.

Clements, Frederick E., and Victor E. Shelford. 1939. Bio-ecology. John Wiley & Sons, Inc., N.Y. 425 p.

Collias, Nicholas E., and Laurence Jahn. 1959. Social behavior and breeding success in Canada geese (*Branta canadensis*) confined under seminatural conditions. Auk 76(4):478–509.

Coues, Elliott. 1874. Birds of the northwest. A handbook of the ornithology of the region drained by the Missouri river and its tributaries. U.S. Dept. Interior. U.S. Government Printing Office, Washington, D.C. 791 p.

Cooke, W. W. 1897. The birds of Colorado. Colo. State Agric. Coll., Agric. Exp. Sta. Bull. 37. Tech. Ser. 2, 143 p.

——— 1898. Further notes on the birds of Colorado. Colo. Agric. Exper. Sta. Bull. 44:148–176.

——— 1900. The birds of Colorado. A second appendix to Bulletin 37, Colo. Agric. Exper. Sta. Bull. 56:179–238.

——— 1906. Distribution and migration of North American ducks, geese and swans. U.S. Biol. Surv. Bull. 26:90 p.

Craighead, Frank C. and John J. Craighead. 1949. Nesting Canada geese on the Upper Snake river. Jour. Wildl. Mgt. 13(1):51–64.

Craighead, John J. and Dwight S. Stockstad. 1964. Breeding age of Canada geese. Jour. Wildl. Mgt. 28(1):57–64.

Critcher, Stuart. 1950. Renal coccidiosis in Pea Island Canada geese. Wildl. in North Carolina 14(11):14–15, 22.

Davie, Oliver. 1898. Nests and eggs of North American birds. 5th Ed. Revised. The Landon Press, Columbus. Pt. 1, 509 p.; Pt. 2, 18 p.; index, xxi p.

Dawson, William Leon. 1923. The birds of California. South Moulton Company. San Diego. Vol. 3:1433–2121.

Delacour, Jean. 1951. Preliminary note on the taxonomy of Canada geese, *Branta canadensis*. Amer. Mus. Novitiates, No. 1537, 10 p.

——— 1954. The waterfowl of the world. Country Life Lmtd., London, Vol. I, 284 p.

Domm, L. V., and E. Taber. 1946. Endocrine factors controlling erythrocyte concentrations in the blood of the domestic fowl. Physiol. Zool. 19(3): 258–281.

Dow, Jay S. 1943. A study of nesting Canada geese in Honey Lake Valley, California. Calif. Fish and Game 29(1):3–18.

Dutcher, William. 1885. The Canada goose. Auk 2(1):111.

Elder, William H. 1946. Age and sex criteria and weights of Canada geese. Jour. Wildl. Mgt. 10(2):93–111.

Errington, Paul L. 1946. Predation and vertebrate populations. Quart. Rev. of Biol. 21(2):144–177; 21(3):221–245.

——— 1962. Muskrat populations. Iowa State Univ. Press, Ames, Iowa. 665 p.

Farr, Marion M. 1952. Tyzzeria sp. from wild geese and a wild duck. Jour. Parasitol. Suppl. 3(4):15.

——— 1953. Three new species of coccidia from the Canada goose, *Branta canadensis* (Linne, 1958) Jour. Wash. Acad. Sci. 43(10):336–340.

——— 1954. Renal coccidiosis of Canada geese. Jour. Parasitol. Suppl. 40(5): 46.

Fenneman, Nevin M. 1931. Physiography of western United States. McGraw-Hill Book Co., N.Y. 534 p.

——— 1938. Physiography of eastern United States. McGraw-Hill Book Co., N.Y. 714 p.

Figgins, J. D. 1920. The status of the subspecific races of *Branta canadensis*. Auk 37(1):94–102.

Flint, Richard F. [Chairman]. 1945. Glacial map of North America. First ed. Geol. Soc. of America, N.Y.

——— 1957. Glacial and pleistocene geology. John Wiley & Sons, Inc., N.Y. 553 p., 5 pl.

Flint, Richard F., Roger B. Cotton, Richard P. Goldthwait and H. B. Willman. 1959. Glacial map of the United States east of the Rocky Mountains. Geol. Soc. of America, N.Y.

Ford, Alice. 1957. The bird biographies of John James Audubon. Macmillan Co., N.Y. 282 p.

Geis, Mary Barraclough. 1956. Productivity of Canada geese in the Flathead Valley, Montana. Jour. Wildl. Mgt. 20(4):409:419.

Gibb, John A. 1961. Bird populations, p. 413–446. *In* A. J. Marshall, Biology and comparative physiology of birds. Academic, N.Y. 2 vols.

Godfrey, W. Earl. 1953. Notes on birds of the area of integration between eastern prairie and forest in Canada. Nat. Mus. Canada. Ann. Rep. 1951–52. Bull. 128:52 p.

Goode, J. Paul. 1943. Goode's school atlas. Physical, political and economic. Rand McNally and Co., Chicago. 286 p.

Goss, N. S. 1891. History of the birds of Kansas. Geo. W. Crane & Co., Topeka, Kan. 693 p.

Gower, W. Carl. 1939. The use of the bursa of Fabricius as an indication of age in game birds. N. Amer. Wildlife Conf. Trans. 4:426–430.

Greenway, James C. 1958. Extinct and vanishing birds of the world. Amer. Committee for Internat. Wild Life Protec., Spec. Publ. 13:518 p.

Haecker, F. W., R. Allyn Moser and Jane B. Swenk. 1945. Check-list of the birds of Nebraska. Nebr. Ornithologists' Union. Reprinted from the Nebraska bird review. Vol. 13, Revised Nov. 1945. 44 p.

Hales, B. J. 1927. Prairie birds. Macmillan Company of Canada (Limited), Toronto.

Hammond, M. C., and G. E. Mann. 1956. Waterfowl nesting islands. Jour. Wildl. Mgt. 20(4):345–352.

Hansen, Henry A., and Wendell H. Oliver. 1951. The status of resident geese in south-central Washington, Spring, 1950. Murrelet 32(1):2–7.

Hanson, Harold C. 1949*a*. Methods to determine age in Canada geese and other waterfowl. Jour. Wildl. Mgt. 13(2):177–183.

——— 1949*b*. Notes on white spotting and other plumage variations in geese. Auk 66(2):164–171.

——— 1950. A morphometrical study of the Canada goose (*Branta canadensis interior*). Todd. Auk 67(4):164–173.

——— 1953. Interfamily dominance in Canada geese. Auk 70(1):11–16.

——— 1956. A three-year survey of *Ornithofilaria* sp. Microfilariae in Canada geese. Jour. Parasitol. 42(5):543.

——— 1958. Studies on the physiology of Canada geese (*Branta canadensis interior*). Ph.D. Thesis. Univ. Illinois. 125 p.

——— 1959. The incubation patch of wild geese: its recognition and significance. Arctic 12(3):139–150.

——— 1962*a*. The dynamics of condition factors in Canada geese and their relation to seasonal stresses. Arctic Institute of North America. Technical paper 12:68 p.

——— 1962*b*. Characters of age, sex, and sexual maturity in Canada geese. Ill. Nat. Hist. Surv. Biol. Notes 49. 15 p.

——— 1962*c*. Some comparative aspects of organ weights in Canada geese (*Branta canadensis interior*). Ill. State Acad. Sci. Trans. 55(1):58–69.

Hanson, Harold C., and Robert H. Smith. 1950. Canada geese of the Mississippi Flyway, with special reference to an Illinois flock. Ill. Nat. Hist. Surv. Bull. 25(3):67–210.

Hanson, Harold C., Norman D. Levine and S. Kantor. 1956. Filariae in a wintering flock of Canada geese. Jour. Wildl. Mgt. 20(1):89–92.

Hanson, Harold C., Paul Queneau and Peter Scott. 1956. The geography, birds, and mammals of the Perry River Region. Arctic Institute of North America. Spec. Pub. 3, 98 p.

Hanson, Harold C. and Campbell Currie. 1957. The kill of wild geese by the natives of the Hudson–James Bay region. Arctic 10(4):211–229.

Hanson, Harold C., Norman D. Levine and Virginia Ivens. 1957. Coccidia (PROTOZOA: Eimeriidae) of North American wild geese and swans. Canad. Jour. Zool. 35(6):715–733.

Hanson, Harold C., and James R. Gilford. 1961. The prevalence of some helminth parasites in Canada geese wintering in southern Illinois. Ill. State Acad. of Sci. Trans. 54(1 and 2):41–53.

Hanson, W. C., and R. L. Browning. 1959. Nesting studies of Canada geese on the Hanford Reservation, 1953–56. Jour. Wildl. Mgt. 23(2):129–137.

Harmon, I. W., E. Ogden and S. F. Cook. 1932. The reservoir function of the spleen in fowls. Amer. Jour. Physiol. 100(1):99–101

Hearne, Samuel. 1795. A journey from Prince of Wales's Fort in Hudson's Bay to the northern ocean, 1769 · 1770 · 1771 · 1772. 1958 Edition, edited by Richard Glover. Macmillan Company of Canada (Limited), Toronto. 301 p.

Hellmayr, Charles E. and Boardman Conover. 1948. Catalogue of birds of the Americas and the adjacent islands: Part 1, No. 2 (Spheniscidae-Anatidae). Field Mus. Nat. Hist. Zool. Ser. 13, Publ. 615, 434 p.

Herman, Carlton M., and Everett E. Wehr. 1954. The occurrence of gizzard worms in Canada geese. Jour. Wildl. Mgt. 18(4):509–513.

Hinde, R. A., and N. Tinbergen. 1958. The comparative study of species—specific behavior. p. 251. *In* Ann Roe and G. G. Simpson (eds.), Behavior and evolution, Yale, New Haven.

Hickey, Joseph J. 1952. Survival studies of banded birds. U.S. Dept. Interior. Fish and Wildl. Service Spec. Sci. Rep.: Wildlife 15:177 p.

——— 1955. Some American population research on gallinaceous birds. 326–396. *In* Albert Wolfson (ed.), Recent studies in avian biology. Univ. of Ill. Press, Urbana.

Hochbaum, H. Albert. 1942. Sex and age determination of waterfowl by cloacal examination. N. Amer. Wildl. Conf. Trans. 7:299–307.

——— 1955. Travels and traditions of waterfowl. Univ. of Minn. Press, Minneapolis. 301 p.

[Holland, Ray P.] 1922. The weight of wild geese. Field and Stream 92(5): 34–35.

Hornaday, William T. 1931. Thirty years war for wildlife. Charles Scribner's Sons, N.Y. 292 p.

Howell, Arthur H. 1911. Birds of Arkansas. U.S. Biol. Serv. Bull. 38:1–100.

Huntington, Dwight W. 1910. Our wild fowl and waders. The Amateur Sportsman Co., N.Y. 207 p.

Irving, Laurence. 1960. Birds of Anaktuvuk Pass, Kobuk, and Old Crow. A study in Arctic adaptation. U.S. Nat. Mus. Bull. 217. 409 p.

Jenkins, Dale W. 1944. Territory as a result of despotism and social organization in geese. Auk 61 (1) :30–47.

Jensen, G. H., and Nolan Nelson. 1948. Goose breeding grounds — Utah and southeastern Idaho. *In* Waterfowl populations and breeding conditions — summer 1948 with notes on woodcock studies. U.S. Fish and Wild Life Service. Spec. Sci. Rep. 60:152–155.

Johansen, Hans. 1945. Om racer af saedgaes. Dansk Ornithologisk Forenings Tidsskrift 39(1):106–127.

Johnson, C. S. 1947. Canada goose management, Seney National Wildlife Refuge. Jour. Wildl. Mgt. 11(1):21–24.

Judd, Elmer T. 1917. List of North Dakota birds. Found in the Big Coulee, Turtle Mountains and Devils Lake region. As noted during the years 1890 to 1896 and verified in the subsequent years to date. Publ. by the author, Cando, North Dakota.

Kalmbach, E. R. 1939. Nesting success: its significance in waterfowl reproduction. N. Amer. Wildl. Conf. Trans. 4:591–604.

Kebbe, Chester E. 1955. Waterfowl breeding ground survey, Oregon, 1954, p. 161–164. *In* Waterfowl populations and breeding conditions — summer 1954. U.S. Fish and Wildl. Service. Spec. Sci. Rep.-Wildl. 27:288.

Kendrew, W. G., and B. W. Currie. 1955. The climate of central Canada. Manitoba, Saskatchewan, Alberta and the districts of MacKenzie and Keewatin. Queens Printer, Ottawa, Canada. 194 p.

Kliph. 1881. The Franklin Club at Reelfoot. Forest and Stream 16:244–245.

Klopman, Robert B. 1958. The nesting of the Canada goose at Dog Lake, Manitoba. Wilson Bull. 70(2):168–183.

Kossack, Charles W. 1947. Incubation temperatures of Canada geese. Jour. Wildl. Mgt. 11(2):119–126.

——— 1950. Breeding habits of Canada geese under refuge conditions. Amer. Midl. Nat. 43(3):627–649.

Krummes, William T. 1941. The muskrat: a factor in waterfowl habitat management. N. Amer. Wildl. Conf. Trans. 5:395–398.

Kumlien, Ludwig, and Ned Hollister. 1903. The birds of Wisconsin. Bull. Wis. Nat. Hist. Soc. 3(1, 2, and 3):143 p. Milwaukee Public Museum, Milwaukee, Wisconsin.

Kuyt, E. 1962. Northward dispersion of banded Canada geese. Canad. Field-Naturalist 76(3):180–181.

Lajeunesse, E. J. 1960. The Windsor Border Region, Canada's southernmost frontier: a collection of documents. The Champlain Society for the Government of Ontario. Univ. Toronto Press, Ont. Series No. 4. 129: 374 p.

Lawrence, A. G. 1925. *In* Chickadee Notes, No. 234. Manitoba Free Press, 17 Sept.

Levine, Norman D. 1952. *Eimeria magnalabia* and *Tyzzeria* sp. (Protozoa: Eimeriidae) from the Canada goose. Cornell Vet. 42(2):247–252.

Levine, Norman D., and Harold C. Hanson. 1953. Blood parasites of the Canada goose. Jour. Wildl. Mgt. 17(2):185–196.

Lincoln, Frederick C. 1925. Notes on the bird life of North Dakota with particular reference to the summer waterfowl. Auk 42(1):50–64.

Lobeck, A. K. 1932. Physiographic diagram of the United States. The Geographical Press. Columbia Univ., N.Y.

Lynch, John J., and J. R. Singleton. 1964. Winter appraisals of annual productivity in geese and other water birds. Wildfowl Trust. 15th Ann. Rep. p. 114–126.

Macoun, John. 1883. Manitoba and the great Northwest. Thomas C. Jack, London. 687 p.

Macoun, John, and James M. Macoun. 1909. Catalogue of Canadian birds. Canada Dept. of Mines, Geol. Surv. Branch, Ottawa. No. 973, 761 p.

Main, Angie Kumlien. 1943. Thure Kumlien, Koshkonong naturalist (II). Wis. Mag. Hist. 27(2):194–220.

Manitoba Dept. of Mines and Resources. 1945. Game birds and animals of Manitoba. Winnipeg. 47 p.

Marshall, A. J. 1961. Biology and comparative physiology of birds. Academic Press, N.Y. 2 vols.

Mayne, R. C. 1862. Four years in British Columbia and Vancouver Island. Account of their forests, rivers, gold fields, and resources for colonization. John Murray, London. 468 p.

Mayr, Ernst. 1951. Speciation in birds. International Ornithological Congress Upsala. p. 91–131.

——— 1958. Behavior and Systematics. *In* Roe and Simpson, *op. cit.*, p. 341–362.

——— 1963. Animal species and evolution. Harvard University Press, Cambridge, Mass. 797 p.

McAtee, W. L. 1944. Popular subspecies. Auk 61(1):135–136.

McEwen, Eoim H. 1958. Observations on the lesser snow goose nesting grounds, Egg River, Banks Island. Canad. Field-Naturalist 72(3): 122–127.

McKinley, Daniel. 1961. History of the Canada goose in Missouri. The Bluebird 28(3): 2–8.

Meister, Waldemar. 1951. Changes in the histological structure of the long-bones of birds during the molt. Anat. Rec. 111(1):1–22.

Mershon, William B. 1923. Recollections of my fifty years hunting and fishing. Stratford, Boston.

——— 1925. Big geese. Field and Stream 30(1):26–27, 63–64.

Miller, A. W., and D. B. Collins. 1953. A nesting study of Canada geese on Tule Lake and Lower Klamath National Wildlife refuges, Siskiyou County, California. Calif. Fish and Game 39(3):385–396.

Miller, Loye. 1924. *Branta dickeyi* from the McKittrick pleistocene (with four drawings). Condor 26(5):178–180.

Miner, Jack. 1923. Jack Miner and the birds. Reilly and Lee Co., Chicago. 176 p.

Mitchell, H. H. 1962. Comparative nutrition of man and domestic animals. Academic Press, N.Y. Vol. I. 701 p.

Moffitt, James. 1931. The status of the Canada goose in California. Calif. Fish and Game 17(1):20–26.

——— 1931*b*. Banding Canada geese in California in 1931. Condor 33(4): 229–237.

Munro, David A. 1960. Factors affecting reproduction of the Canada goose. XII International Ornithological Congress. Helsinki 2:542–556.

Naylor, A. E. 1953. Production of the Canada goose on Honey Lake Refuge, Lassen County, California. Calif. Fish and Game 39(1):83–94.

Naylor, A. E., and E. G. Hunt. 1954. A nesting study and population survey of Canada geese on the Susan river, Lassen county, California. Calif. Fish and Game 40(1):5–16.

Nelson, E. W. 1876. Birds of northeastern Illinois. Bull. of the Essex Institute 8:138 p.

Nelson, Harvey K. 1963. Restoration of breeding Canada goose flocks in the north central states. N. Amer. Wildl. and Nat. Resources Conf. Trans. 28:133–150.

Nelson, Urban C., and Henry A. Hansen. 1959. The cackling goose — its migration and management. N. Amer. Wildl. Conf. Trans. 24:174–186.

Newhouse, Sewell. 1874. The trapper's guide. 6th Edition. New York, N.Y. 118 p.

Over, William H., and Craig S. Thomas. 1921. Birds of South Dakota. S.D. Geol. and Nat. Hist. Surv. Bull. Ser. 21(9):142 p.

Parmelee, David F., and S. P. Macdonald. 1960. The birds of west central Ellesmere Island and adjacent areas. Nat. Mus. Canada Bull. 169:1–103.

Phillips, John C., and Frederick C. Lincoln. 1930. American waterfowl. Their present situation and the outlook for their future. Houghton Mifflin Co., Boston. 312 p.

Pirnie, Miles D. 1938. Restocking of the Canada goose successful in southern Michigan. N. Amer. Wildl. Conf. Trans. 3:624–627.

Preble, Edward A. 1902. A biological investigation of the Hudson Bay region. North American Fauna 22:1–140. U.S. Dept. Agric. Div. Biol. Surv. U.S. Government Printing Office, Washington, D.C.

Raisz, Erwin. 1954. Land form of the United States (Map). *In* W. W. Atwood, Physiographic provinces of North America. Ginn, Boston.

Randall, Thomas E. 1962. Birds of the Kazan Lake region, Saskatchewan. The Blue Jay 20(2):60–72.

Rensch, Bernhard. 1956. Increase of learning capability with increase of brain size. Amer. Naturalist 90(851):81–95.

——— 1959. Trends toward progress of brains and sense organs. Cold Spring Harbor symposia on quantitative biology 24:291–303.

Riddle, O. 1938. The changing organism. *In* Cooperation in research. Carnegie Institution of Washington. Publ. 501:259–273.

Ritchie, William A. 1944. The pre-Iroquoian occupations of New York state. Rochester Museum of Arts and Sciences. Memoir No. 1.

Roberts, Thomas S. 1932. The birds of Minnesota. Univ. of Minn. Press, Minneapolis. Vol. I. 691 p.

Roe, Anne and George Gaylord Simpson. 1958. Behavior and evolution. Yale University Press, New Haven. 557 p.

Rowan, William. 1922. Some bird notes from Indian Bay, Man. Auk 39(2): 224–232.

Rowe, J. S. 1959. Forest regions of Canada. Canada Dept. of Northern Affairs and Nat. Resources. Forestry Branch. Bull. 123. 71 p.

Salt, Ray W., and A. L. Wilk. 1958. The birds of Alberta. Alberta Dept. of Economic Affairs. Queen's Printer, Edmonton. 511 p.

Schantz, H. L., and R. Zon. 1924. Natural vegetation. Atlas of Amer. Agric. U.S. Dept. of Agric. Physical basis of agric. Sec. E:1–29.

Schiøler, E. Lehn. 1924. Om de skandinaviske Ænder, deres Dragtskifte og Traek. Dansk Ornithologisk Forenings Tidsskrift 18:85–95.

Schorger, A. W. 1944. Wisconsin's greatest pioneer zoologist, Philo Romayne Hoy. Passenger Pigeon 6(2):55–59.

Sclater, William Lutley. 1912. A history of the birds of Colorado. Witherby & Co., London. 576 p.

Scott, Peter, Hugh Boyd and J. L. Sladen. 1955. The wildfowl trust's second expedition to central Iceland, 1953. The Wildfowl Trust Seventh Ann. Rep. 1953–54:63–98.

Scott, Peter, James Fisher, and Finnur Gudmundsson. 1951–52. The Severn Wildfowl Trust expedition to central Iceland. Severn Wildfowl Trust Fifth Ann. Rep. 1951–52:78–115.

Selwyn, Alfred, R. C. 1875. Observations in the N.W.T., Fort Garry to Rocky Mt. House and return. Report of progress, 1873–74. Geol. Survey of Canada. Montreal [now Ottawa]. p. 17–62.

Shortt, T. M., and Sam Waller. 1937. The birds of the Lake St. Martin region, Manitoba. Roy. Ont. Mus. Zool. Contrib. 10:1–51.

Society of Amer. Foresters. 1954. Forest cover types of North America (Exclusive of Mexico). Washington, D.C. 67 p.

Steel, Paul E., Paul D. Dalke, and Elwood G. Bizeau. 1957. Canada goose production at Gray's Lake, Idaho, 1949–51. Jour. Wildl. Mgt. 21(1): 38–41.

Sterling, R. Thomas. 1963. Wascana goose summers on the Arctic prairie. Blue Jay 21(4).

Taverner, P. A. 1931. A study of *Branta canadensis* (Linnaeus), the Canada goose. Nat. Mus. Canada Bull. 67. 28–40.

Thacker, Henry. 1874. Muskrat hunting. (*In* Newhouse, *op. cit.*, 1874.)

Thompson, Ernest E. 1891. The birds of Manitoba. Proc. U.S. Nat. Mus. 13(841):457–643.

United States Department of Agriculture. 1941. Climate and man. Yearbook of Agriculture. U.S. Government Printing Office, Washington, D.C. 1248 p.

——— 1938. Soils of the United States. Yearbook of Agriculture. p. 1019–1161. U.S. Government Printing Office, Washington, D.C., 1232 p.

Vaiden, M. G. 1964. Notes on Mississippi birds. The Mississippi kite. Occas. Papers Miss. Nat. Club. 1(8):1–2.

Vaught, Richard W. 1960. Geese by the tub-full. Missouri Conservationist 21(7):6–10.

Visher, Stephen Sargent. 1954. Climatic atlas of the United States. Harvard University Press, Cambridge. 403 p.

Wehr, Everett E., and Carlton M. Herman. 1954. Age as a factor in acquisition of parasites by Canada geese. Jour. Wildl. Mgt. 18(2):239–247.

Wetmore, Alexander. 1958. Miscellaneous notes on fossil birds. Smithsonian Misc. Coll. 135(8):1–11.

"Widgeon" [Cornelius Ackerson]. 1922. Wild-fowling days in Kansas. Forest and Stream 92(1):22–23, 42–44.

Wiedman, Otto. 1907. A preliminary catalog of the birds of Missouri. Trans. Acad. of Sci. of St. Louis 17(1):1–288.

Williams, Cecil S., and Marcus C. Nelson. 1943. Canada goose nests and eggs. Auk 60(3):341–345.

Wilson, J. T. [Chairman]. 1958. Glacial map of Canada. Geological Association of Canada.

Wood, Jack S. 1964. Normal development and causes of reproductive failure in Canada geese. Jour. Wildl. Mgt. 28(2):197–208.

Wood, Norman A. 1923. A preliminary survey of the bird life of North Dakota. Univ. Mich. Mus. Zool. Misc. Publ. No. 10. 96 p.

Woodruff, Frank Morley. 1907. The birds of the Chicago area. Chicago Acad. of Sci. Bull. No. 6. 221 p.

Yocum, Charles F. 1952. Techniques used to increase nesting of Canada geese. Jour. Wildl. Mgt. 16(4):425–428.

——— 1956. Man-made homes for Canada geese. Audubon Magazine 58(3): 106–109, 127.

——— 1962. History of the Great Basin Canada goose in the Pacific Northwest and adjacent areas. Murrelet 43(1):1–9.

INDEX

Most references to flocks, populations, or data pertaining to them, their characteristics, and areas of use have been consolidated under specific locality entries or under the names of the refuges chiefly involved.